Fundamentals of Astrophysics

Fundamentals of Astrophysics

Mervin Williamson

MURPHY & MOORE
www.murphy-moorepublishing.com

Fundamentals of Astrophysics
Mervin Williamson
ISBN: 978-1-63987-243-5 (Hardback)

MURPHY & MOORE

Published by Murphy & Moore Publishing,
1 Rockefeller Plaza,
New York City, NY 10020, USA

Cataloging-in-Publication Data

Fundamentals of astrophysics / Mervin Williamson.
p. cm.
Includes bibliographical references and index.
ISBN 978-1-63987-243-5
1. Astrophysics. 2. Cosmic physics. 3. Astronomy. 4. Physics. I. Williamson, Mervin.
QB461 .F68 2022
523.01--dc23

Table of Contents

Preface

Astrophysics is a branch of science which utilizes the methods and principles of physics to study the astronomical objects and phenomena. It is a multidisciplinary field which incorporates concepts and methods from many disciplines such as classical mechanics, electromagnetism, statistical mechanics, thermodynamics, quantum mechanics, relativity, nuclear and particle physics, and atomic and molecular physics. This subject focuses on the study the Sun, stars, galaxies, extrasolar planets, the interstellar medium and cosmic wave background. The emissions from these objects are analyzed across the electromagnetic spectrum for characterization of properties like luminosity, temperature, density, chemical composition, etc. Astronomical research often involves a substantial amount of work in the field of theoretical and observational physics. Some areas of study under this discipline include determining properties of dark matter, dark energy, black holes, other celestial bodies and the origin of the universe. Observational astrophysics is a sub branch of astrophysics which focuses on the techniques to record and interpret data. This book is compiled in such a manner, that it will provide in-depth knowledge about the theory and practice of astrophysics. The various sub-fields of astrophysics along with technological progress that have future implications are glanced at in it. This book will provide comprehensive knowledge to the readers.

A foreword of all chapters of the book is provided below:

Chapter 1 - Astrophysics is a branch of science that deals with study of celestial bodies. It applies the laws of physics to explain the birth, life and death of stars, planets, galaxies, nebulae and other objects in the universe. Electromagnetic spectrum, blackbody radiation, luminosity, flux, Boltzmann equation, etc. are a few terms that fall under this field. This chapter will serve as an introduction to astrophysics; **Chapter 2** - There are many branches of Astrophysics such as observational astrophysics, theoretical astrophysics, nuclear astrophysics, computational astrophysics, etc. This chapter deals with each branch and its specific subject matter in a clear and concise manner for the benefit of the reader; **Chapter 3** - Astrophysics has numerous theories associated with it. These include the Big Bang which proposes the creation of the universe and its ever expanding nature and the Partial Impact Theory which describes the creation of a third star through the partial collision of two stars fall within its aspect. This chapter aims to shed light on the various astrophysics theories to provide a better understanding; **Chapter 4** - Galaxies are the vast cosmic islands of stars, gases, dust and dark matter that span the vast universe. All aspects related to galaxies are addressed under the astrophysics of galaxies. The formation of galaxies, their structures, the different types of galaxies and the various dynamics that act upon them, including our very own Milky Way, are the topics that are discussed in-depth in this chapter; **Chapter 5** - Stars are the celestial bodies made out of hydrogen and helium that produce light and heat from the continuous nuclear reaction taking place at their cores. Star formation and stellar evolution are topics actively pursued by scientists in order to understand these astronomical objects better, given that our sun is also a star. This chapter covers all matters related to stars and stellar astrophysics comprehensively; **Chapter 6** - Solar astrophysics is a branch of astrophysics that specializes in the study of the sun. A detailed understanding of the sun is crucial before contemplating the formation and evolution of the Solar System, the Ring System, the Planetary System and other similar alien systems which may be discovered in the near future. This chapter provides a complete and detailed examination of solar astrophysics for a clearer understanding.

I would like to thank the entire editorial team who made sincere efforts for this book and my family who supported me in my efforts of working on this book. I take this opportunity to thank all those who have been a guiding force throughout my life.

Mervin Williamson

1

Basics of Astrophysics

Astrophysics is a branch of science that deals with study of celestial bodies. It applies the laws of physics to explain the birth, life and death of stars, planets, galaxies, nebulae and other objects in the universe. Electromagnetic spectrum, blackbody radiation, luminosity, flux, Boltzmann equation, etc. are a few terms that fall under this field. This chapter will serve as an introduction astropohysics.

Astrophysics is the science of physical processes in the cosmos. It uses data gathered by astronomers using telescopes on Earth and in space – combined with the laws and theories of physics – in order to interpret the universe around us. If astronomy asks what and where, astrophysics asks how and why. A sister science – planetary science – studies the planets in our solar system and distant solar systems in our Milky Way galaxy. Another sister science – cosmology – studies external galaxies and voids, and the large-scale structure and history of the universe.

For example, an astronomer might spend nights at the telescope gathering data on a star. Putting on his or her astrophysicist hat – and depending on which instruments were used in conjunction with the telescope (photometers or spectrometers, for example) – that scientist would then turn to the laws of physics to understand how that star produces its energy, whether it has a companion (or perhaps planets, or perhaps an encircling disk), and how the star moves through space.

The history of astrophysics really begins in Europe with the Renaissance, when astronomy threw off the shackles of eons-old myths and became a true science. In Florence, in 1610, Galileo Galilei became one of the first to turn the telescope to the sky, revealing a universe at odds with the restrictive teachings of the Church. By 1633, this dangerous contradiction saw Galileo tried by the Roman Inquisition and confined to house arrest for the rest of his life until his death in 1642.

In England in 1665, Isaac Newton, having fled Cambridge University to escape the Great Plague, settled back in the town of his birth, the rural community of Woolsthorpe, Lincolnshire. Over the next several years, he developed calculus and revolutionary ideas about light. In 1687, he published his Law of Universal Gravitation, stating that every particle attracts every other particle in the universe with a force that is directly proportional to the product of their masses and inversely proportional to the square of the distance between them. For the first time, scientists could calculate the force of attraction between objects in space. This was a giant step forward in scientists' ability to understand the heavens.

Remarkable discoveries have been made in astrophysics in recent time ranging from the Big Bang and the early Universe, to the Cosmic Microwave Background and its fluctuation spectrum, to measurements of the large-scale structure in the Universe, the existence of Dark Matter and Dark Energy, the discovery of Supermassive Black Holes, and the discovery of planets around other stars. Observations needed to probe the universe and answer questions about Universe are carried out mostly with telescopes, not only the familiar ones sensitive to optical light rays, but also with instruments designed to receive radio waves, X-rays, and Gamma-rays. Within the solar system, astronomers use space probes. The vast amount of observational detail obtained with these techniques is then interpreted by means of the basic laws of physics. Especially in recent decades, the new tools of radio telescopes on the ground and X-ray, optical, and ultraviolet telescopes in space have permitted us to make the startling discoveries about the heavens. In addition, we now know, for example, of dense stars that consist almost entirely of neutrons, with the same amount of material as in the Sun compressed into a sphere only a few miles in diameter, with a resultant density of millions of tons packed into each cubic inch. We find even smaller, more massive, objects -- black holes -- whose gravitational attraction is so great that any light waves from the surface cannot escape but are attracted back. We find that most galaxies contain a supermassive black hole, of many millions or even billions solar masses, at their cores. Gigantic explosions of stars within individual galaxies -- supernovae and gamma-ray bursts -- have been found to radiate as much light as billions of suns. Such explosions have been detected in systems as far out as nearly the edge of the accessible Universe, where stellar systems are moving away from us at close to the speed of light, and from which the light rays we now see were emitted billions of years ago when the Universe was much younger. The Cosmic Microwave Background radiation -- a 3K degree radiation that is a remnant of the hot Big Bang some 14 billion years ago. This radiation is remarkably uniform. However, on top of this highly uniform distribution, the tiny fluctuations that provided the seeds for galaxy and structure formation in the early universe have been detected and carefully mapped, a discovery of great importance for understanding how the structure we see today formed. While such discoveries are fascinating in their own right, they cast light on the fundamental questions that people have been asking since the dawn of mankind about the hidden nature of our Universe.

Electromagnetic Spectrum

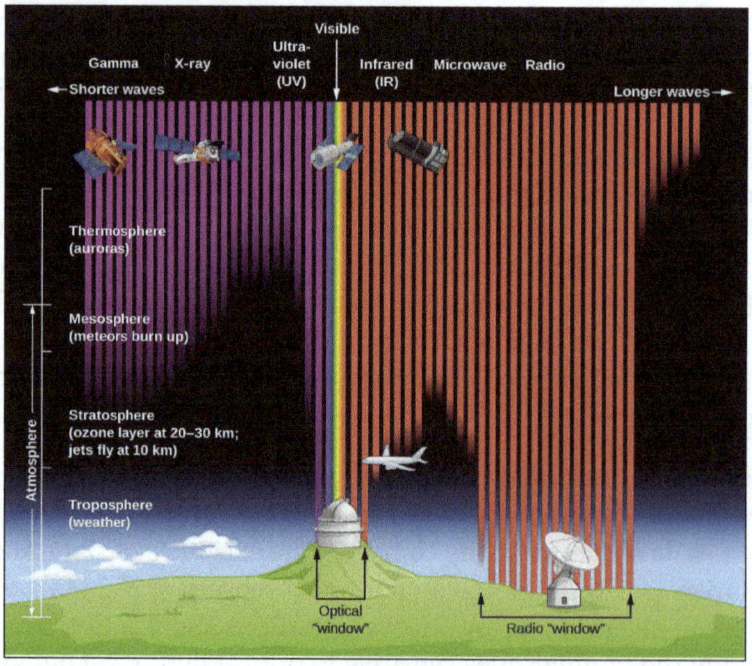

Radiation and Earth's Atmosphere. This figure shows the bands of the electromagnetic spectrum.

and how well Earth's atmosphere transmits them. Note that high-frequency waves from space do not make it to the surface and must therefore be observed from space. Some infrared and microwaves are absorbed by water and thus are best observed from high altitudes. Low-frequency radio waves are blocked by Earth's ionosphere.

Objects in the universe send out an enormous range of electromagnetic radiation. Scientists call this range the electromagnetic spectrum, which they have divided into a number of categories. The spectrum is shown in Figure, with some information about the waves in each part or band.

Types of Electromagnetic Radiation

Electromagnetic radiation with the shortest wavelengths, no longer than 0.01 nanometer, is categorized as gamma rays (1 nanometer = 10^{-9} meters; see Units Used in Science). The name gamma comes from the third letter of the Greek alphabet: gamma rays were the third kind of radiation discovered coming from radioactive atoms when physicists first investigated their behavior. Because gamma rays carry a lot of energy, they can be dangerous for living tissues. Gamma radiation is generated deep in the interior of stars, as well as by some of the most violent phenomena in the universe, such as the deaths of stars and the merging of stellar corpses. Gamma rays coming to Earth are absorbed by our atmosphere before they reach the ground (which is a good thing for our health); thus, they can only be studied using instruments in space.

Electromagnetic radiation with wavelengths between 0.01 nanometer and 20 nanometers is referred to as X-rays. Being more energetic than visible light, X-rays are able to penetrate soft tissues but not bones, and so allow us to make images of the shadows of the bones inside us. While X-rays can penetrate a short length of human flesh, they are stopped by the large numbers of atoms in Earth's atmosphere with which they interact. Thus, X-ray astronomy (like gamma-ray astronomy) could not develop until we invented ways of sending instruments above our atmosphere.

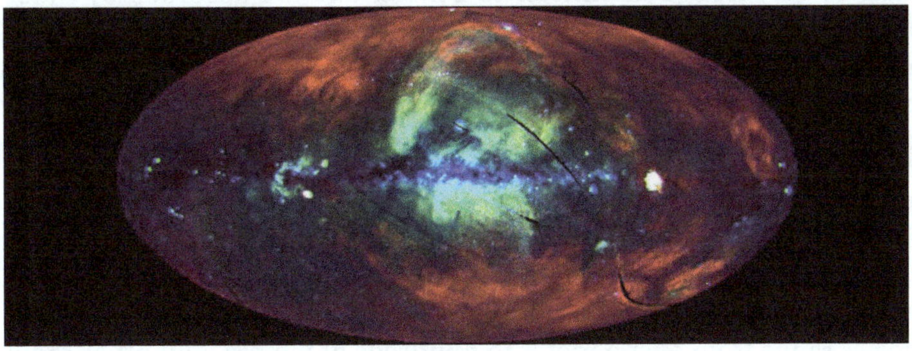

X-Ray Sky.

This is a map of the sky tuned to certain types of X-rays (seen from above Earth's atmosphere). The map tilts the sky so that the disk of our Milky Way Galaxy runs across its center. It was constructed and artificially colored from data gathered by the European ROSAT satellite. Each color (red, yellow, and blue) shows X-rays of different frequencies or energies. For example, red outlines the glow from a hot local bubble of gas all around us, blown by one or more exploding stars in our cosmic vicinity. Yellow and blue show more distant sources of X-rays, such as remnants of other exploded stars or the active center of our Galaxy (in the middle of the picture).

Radiation intermediate between X-rays and visible light is ultraviolet (meaning higher energy than violet). Outside the world of science, ultraviolet light is sometimes called "black light" because our eyes cannot see it. Ultraviolet radiation is mostly blocked by the ozone layer of Earth's atmosphere, but a small fraction of ultraviolet rays from our Sun do penetrate to cause sunburn or, in extreme cases of overexposure, skin cancer in human beings. Ultraviolet astronomy is also best done from space.

Electromagnetic radiation with wavelengths between roughly 400 and 700 nm is called visible light because these are the waves that human vision can perceive. This is also the band of the electromagnetic spectrum that most readily reaches Earth's surface. These two observations are not coincidental: human eyes evolved to see the kinds of waves that arrive from the Sun most effectively. Visible light penetrates Earth's atmosphere effectively, except when it is temporarily blocked by clouds.

Between visible light and radio waves are the wavelengths of infrared or heat radiation. Astronomer William Herschel first discovered infrared in 1800 while trying to measure the temperatures of different colors of sunlight spread out into a spectrum. He noticed that when he accidently positioned his thermometer beyond the reddest color, it still registered heating due to some invisible energy coming from the Sun. This was the first hint about the existence of the other (invisible) bands of the electromagnetic spectrum, although it would take many decades for our full understanding to develop.

A heat lamp radiates mostly infrared radiation, and the nerve endings in our skin are sensitive to this band of the electromagnetic spectrum. Infrared waves are absorbed by water and carbon dioxide molecules, which are more concentrated low in Earth's atmosphere. For this reason, infrared astronomy is best done from high mountaintops, high-flying airplanes, and spacecraft.

After infrared comes the familiar microwave, used in short-wave communication and microwave ovens. (Wavelengths vary from 1 millimeter to 1 meter and are absorbed by water vapor, which makes them effective in heating foods.) The "micro-" prefix refers to the fact that microwaves are small in comparison to radio waves, the next on the spectrum. You may remember that tea—which is full of water—heats up quickly in your microwave oven, while a ceramic cup—from which water has been removed by baking—stays cool in comparison.

All electromagnetic waves longer than microwaves are called radio waves, but this is so broad a category that we generally divide it into several subsections. Among the most familiar of these are radar waves, which are used in radar guns by traffic officers to determine vehicle speeds, and AM radio waves, which were the first to be developed for broadcasting. The wavelengths of these different categories range from over a meter to hundreds of meters, and other radio radiation can have wavelengths as long as several kilometers.

With such a wide range of wavelengths, not all radio waves interact with Earth's atmosphere in the same way. FM and TV waves are not absorbed and can travel easily through our atmosphere. AM radio waves are absorbed or reflected by a layer in Earth's atmosphere called the ionosphere (the ionosphere is a layer of charged particles at the top of our atmosphere, produced by interactions with sunlight and charged particles that are ejected from the Sun).

We hope this brief survey has left you with one strong impression: although visible light is what most people associate with astronomy, the light that our eyes can see is only a tiny fraction of the broad range of waves generated in the universe. Today, we understand that judging some astronomical phenomenon by using only the light we can see is like hiding under the table at a big dinner party and judging all the guests by nothing but their shoes. There's a lot more to each person than meets our eye under the table. It is very important for those who study astronomy today to avoid being "visible light chauvinists"—to respect only the information seen by their eyes while ignoring the information gathered by instruments sensitive to other bands of the electromagnetic spectrum.

Table summarizes the bands of the electromagnetic spectrum and indicates the temperatures and typical astronomical objects that emit each kind of electromagnetic radiation. While at first, some of the types of radiation listed in the table may seem unfamiliar, you will get to know them better as your astronomy course continues. You can return to this table as you learn more about the types of objects astronomers study.

Table: Types of Electromagnetic Radiation.

Type of Radiation	Wavelength Range (nm)	Radiated by Objects at This Temperature	Typical Sources
Gamma rays	Less than 0.01	More than 10^8 K	Produced in nuclear reactions; require very high-energy processes.
X-rays	0.01–20	10^6–10^8 K	Gas in clusters of galaxies, supernova remnants, solar corona.
Ultraviolet	20–400	10^4–10^6 K	Supernova remnants, very hot stars.
Visible	400–700	10^3–10^4 K	Stars
Infrared	10^3–10^6	10–10^3 K	Cool clouds of dust and gas, planets, moons.
Microwave	10^6–10^9	Less than 10 K	Active galaxies, pulsars, cosmic background radiation.
Radio	More than 10^9	Less than 10 K	Supernova remnants, pulsars, cold gas.

Radiation and Temperature

Some astronomical objects emit mostly infrared radiation, others mostly visible light, and still others mostly ultraviolet radiation. What determines the type of electromagnetic radiation emitted by the Sun, stars, and other dense astronomical objects? The answer often turns out to be their temperature.

At the microscopic level, everything in nature is in motion. A solid is composed of molecules and atoms in continuous vibration: they move back and forth in place, but their motion is much too small for our eyes to make out. A gas consists of atoms and/or molecules that are flying about freely at high speed, continually bumping into one another and bombarding the surrounding matter. The hotter the solid or gas, the more rapid the motion of its molecules or atoms. The temperature of something is thus a measure of the average motion energy of the particles that make it up.

This motion at the microscopic level is responsible for much of the electromagnetic radiation on Earth and in the universe. As atoms and molecules move about and collide, or vibrate in place, their electrons give off electromagnetic radiation. The characteristics of this radiation are determined by the temperature of those atoms and molecules. In a hot material, for example, the individual particles vibrate in place or move rapidly from collisions, so the emitted waves are, on average, more energetic. And recall that higher energy waves have a higher frequency. In very cool material, the particles have low-energy atomic and molecular motions and thus generate lower-energy waves.

Blackbody Radiation

To understand, in more quantitative detail, the relationship between temperature and electromagnetic radiation, we imagine an idealized object called a blackbody. Such an object (unlike your sweater or your astronomy instructor's head) does not reflect or scatter any radiation, but absorbs all the electromagnetic energy that falls onto it. The energy that is absorbed causes the atoms and molecules in it to vibrate or move around at increasing speeds. As it gets hotter, this object will radiate electromagnetic waves until absorption and radiation are in balance.

Kirchoff's law requires that every black body has emissivity e_v=1 at all frequencies as well, so the radiated spectrum B_v(T) of any black body at temperature T is the same as the equilibrium spectrum of radiation inside a cavity of temperature T, even if the walls of the cavity are not black. Thus the intensity and spectrum of blackbody radiation depends only on the temperature of the black body or cavity. The same is true for the electrical noise generated by a warm resistor, a device that dissipates electrical energy, and which plays an important role in radio astronomy. You may already be familiar with the standard derivations of the Rayleigh-Jeans and Planck radiation laws, but they are worth repeating because blackbody radiation is so fundamental and because their one-dimensional analogs yield the spectrum of electrical noise generated by a resistor.

The Rayleigh-Jeans Law

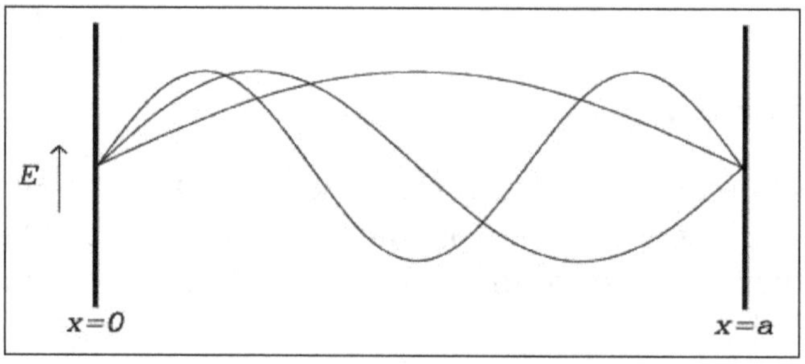

Standing waves corresponding to $n\lambda_x / 2$ =a for n=1 2 3. Abscissa: x-axis of the cavity, bounded by walls at x=0 and x=a. Ordinate: Electric field strengths of the longest standing waves satisfying the boundary condition E=0 at the walls.

Consider a large (meaning side length $a \gg \lambda$, where λ is the longest wavelength of interest) cubical cavity filled with radiation in thermodynamic equilibrium. The purpose of the cavity is to generate radiation and confine the radiation long enough for it to reach equilibrium. The radiation must be generated by thermal accelerations of charged particles in the walls of any cavity with T>0, but the equilibrium radiation is otherwise independent of the wall material. We only require that the walls have nonzero conductivity because walls of zero conductivity would be transparent; they would generate no currents in response to the electric fields of incoming radiation and the radiation would pass through unaffected. For walls having nonzero conductivity, the transverse electric field strength at the walls is E=0 because modes with $E \neq 0$ at the walls are lossy. Only those standing waves with E=0 at the walls will persist after some time $t \gg a/c$.

We can enumerate all possible standing-wave modes in the cavity. For example, consider all standing waves whose wave normals point in the x direction.

The boundary conditions E=0 at x=0 and at x=a mean that only the waves having the discrete wavelengths,

$$\frac{\lambda_X}{2}=a, \quad \frac{2\lambda_X}{2}=a \quad \frac{3\lambda_X}{3}=a,...,$$

can have nonzero amplitudes. Thus,

$$\frac{n\lambda_X}{2}=a,$$

where n=1, 2, 3, ...Similarly,

$$\frac{n_X\lambda_X}{2}=\frac{n_Y\lambda_Y}{2}=\frac{n_Z\lambda_Z}{2}=a$$

for wave normals in the xyz directions. What about a wave whose normal is in some arbitrary direction? Let α, β, γ be the angles between the wave normal and the x,y,z axes, respectively.

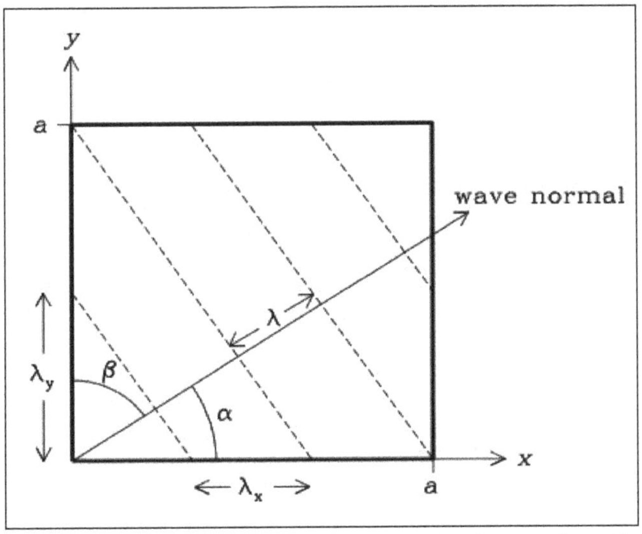

This two-dimensional figure illustrates standing waves propagating in a cavity with wave normals at angles α and β from the x and y axes, respectively. Examples of wave nodes, where E=0, are indicated by dashed lines for the case n_x=3, n_y=2.

From the figure it is clear that,

$$\text{ë}=\text{ë}_X \cos\alpha$$

where ë is the wavelength measured in the direction of the wave normal and $\text{ë}_X \geq \text{ë}$ is the spacing between the wave nodes measured along the x-axis. Thus,

$$\text{ë}_X = \frac{\text{ë}}{\cos\alpha}$$

Similarly,

$$\text{ë}_Y = \frac{\text{ë}}{\cos\beta} \quad and \quad \text{ë}_Z = \frac{\text{ë}}{\cos\lambda}$$

Our boundary conditions become,

$$n_X = \frac{2a}{\text{ë}_X}, \quad n_Y = \frac{2a}{\text{ë}_Y}, \quad n_Z = \frac{2a}{\text{ë}_Z}$$

Squaring and summing the last three equations gives,

$$n_X^2 + n_y^2 + n_Z^2 = (\frac{2a}{\text{ë}})^2, cos^2\alpha + cos^2\beta + cos^2\lambda$$

The Pythagorean Theorem implies $(cos^2\alpha + cos^2\beta + cos^2\lambda)=1$ so,

$$n_X^2 + n_y^2 + n_Z^2 = (\frac{2a}{\text{ë}})^2$$

The permitted frequencies $\upsilon = c / \lambda$ of these standing waves are:

$$\upsilon = \frac{c}{2a}\sqrt{n_X^2 + n_Y^2 + n_Z^2}$$

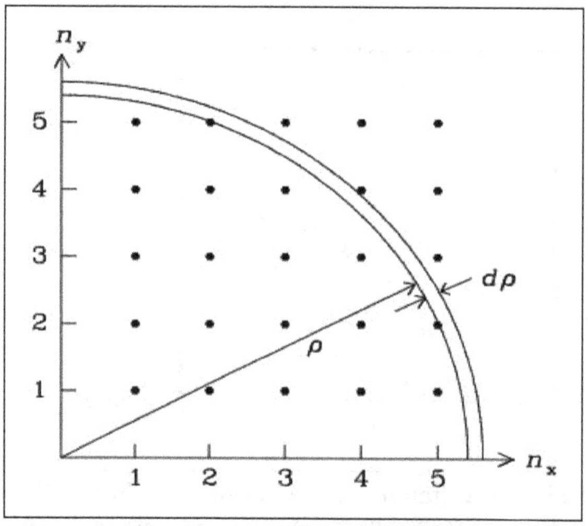

The (x,y) plane in the imaginary space whose axes are (n_x, n_y, n_z). Permitted standing waves in the (n_x, n_y) plane are indicated by dots at positive integer values of these axes.

for all positive integers n_x, n_y, and n_z. The permitted standing-wave modes can be represented as a lattice in the positive octant of the space whose axes are n_x, n_y, and n_z. Each point of the lattice represents one possible mode of equilibrium cavity radiation. The space density of points in this lattice is unity, so the number of points in any volume is equal to that volume.

Let be the radial coordinate in (n_x, n_y, n_z)-space. Then,

$$p^2 = n_X^2 + n_Y^2 + n_Z^2$$

And

$$\upsilon = \frac{c}{\lambda} = \frac{cp}{2a}$$

The number N(v)dv of independent modes having frequencies in the range v to v +d equals the volume of the spherical octant shell between p and p +dp, multiplied by 2 to account for the two independent polarizations of electromagnetic radiation:

$$N(\upsilon)d\upsilon = \frac{4\pi p^2 dp}{8} \times 2$$

$$N(\upsilon)d\upsilon = \pi \left(\frac{2a\upsilon}{c}\right)^2 \frac{2a}{c} d\upsilon$$

In thermodynamic equilibrium at temperature T, equipartition of energy implies that each mode has average energy {E}=kT according to the classical Boltzmann law (but not according to quantum mechanics). If we proceed using the classical kT for the average energy per mode, the spectral energy density $u_v(T)$ of cavity radiation in the frequency range v to v +dv is the total energy of all modes in that frequency range divided by the volume a^3 of the cavity:

$$u_\upsilon(T)d\upsilon = \frac{N(\upsilon)d\upsilon}{a^3} kT$$

$$u_\upsilon(T) = \frac{8\pi a^3}{a^3} \frac{\upsilon^2}{c^3} kT = 8\pi kT \frac{\upsilon^3}{c^3}$$

The spectral energy density of radiation (blackbody or not) is spectral energy per unit volume. It equals the total flow of spectral power per unit area divided by the flow speed c:

$$u_\upsilon = \frac{1}{c} \int I_\upsilon d\Omega$$

Calling the specific intensity of blackbody radiation B_v and making use of the fact the blackbody radiation is isotropic, we get for blackbody radiation:

$$u_\upsilon = \frac{1}{c} \int_{4\pi} B_\upsilon d\Omega = \frac{4\pi}{c} B_\upsilon$$

Thus,

$$\frac{4\pi}{c}B_{\upsilon} = \frac{8\pi kT\upsilon^2}{c^3}$$

And we obtain the Rayleigh-Jeans law for the spectral brightness of blackbody radiation,

$$B_{\upsilon} = \frac{2kT\upsilon^2}{c^2}$$

That is valid only in the low-frequency limit $h\upsilon \ll kT$.

Notice that:

- Brightness is proportional to frequency squared because the volume of a spherical shell in three dimensions is proportional to frequency squared.
- We assumed all modes have $\langle E \rangle = kT$, the classical assumption that breaks down at high frequencies.
- B_v diverges at high. This is called the "ultraviolet catastrophe".
- B_v is independent of direction.
- This result derived for a cubical cavity applies to equilibrium radiation in a cavity of any shape or (large) size, which you can demonstrate by a thought experiment in which a cubical cavity is connected through a small hole to the other cavity.

The Planck Radiation Law

The only flaw in the derivation of the Rayleigh-Jeans law is the classical assumption that each radiation mode can have any energy E. Then the continuous Boltzmann probability distribution.

$$P(E) \propto \exp\left(\frac{-E}{kT}\right)$$

implies an average energy per mode,

$$\langle E \rangle = \frac{\int_0^\infty EP(E)dE}{\int_0^\infty P(E)dE} = kT$$

Planck postulated that possible mode energies are not continuously distributed, but rather they are quantized and must satisfy,

$$E = nh\upsilon, \qquad n = 1, 2, 3, \ldots$$

Where $h \approx 6.63 \times 10^{-27}$ erg s is Planck's constant and n is the number of photons (particles of light each having energy E=hv) in that mode. Then,

$$P(E) = P(nh\upsilon) \propto \exp\left(\frac{-nh\upsilon}{kT}\right)$$

And the average energy per mode is calculated by summing over only the discrete energies permitted instead of integrating over all energies.

$$\langle E \rangle = \frac{\sum_{n=0}^{\infty} nh\upsilon P(nh\upsilon)}{\sum_{n=0}^{\infty} P(nh\upsilon)} = \frac{\sum_{n=0}^{\infty} nh\upsilon \exp\left(\frac{-nh\upsilon}{kT}\right)}{\sum_{n=0}^{\infty} \exp\left(\frac{-nh\upsilon}{kT}\right)}$$

Evaluating this sum yields,

$$\langle E \rangle = \frac{h\upsilon}{\exp\left(\frac{h\upsilon}{kT}\right) - 1} = kT\left[\frac{h\upsilon/(kT)}{\exp[h\upsilon/(kT) - 1]}\right]$$

Thus the correct radiation law becomes,

$$B_\upsilon = \frac{2kT\upsilon^2}{c^2}\left[\frac{h\upsilon/(kT)}{\exp[h\upsilon/(kT) - 1]}\right]$$

where the first factor is the Rayleigh-Jeans law and the quantity in brackets is the quantum correction factor. Planck's law for blackbody radiation is usually written in the form,

$$B_\upsilon = \frac{2h\upsilon^3}{c^2} \frac{1}{\exp\left(\frac{h\upsilon}{kT}\right) - 1}$$

This eliminates the ultraviolet catastrophe because integrating over all frequencies gives a finite integrated brightness a blackbody radiator at temperature T:

$$B(T) = \int_0^{\infty} B_\upsilon(T) d\upsilon = \frac{\sigma T^4}{\pi}$$

Where,

$$\sigma \equiv \frac{2\pi^5 k^4}{15 c^2 h^3} \approx 5.67 \times 10^{-5} \frac{erg}{cm^2 s K^4 sr}$$

is called the Stefan-Boltzmann constant. The spectral energy density per unit solid angle of radiation is just the specific intensity I_v divided by the flow rate c. For isotropic radiation, the spectral energy density is,

$$u_v = 4\pi I_v / c$$

so the total radiation energy density $u = \int_0^\infty u_v dv$ of blackbody radiation is,

$$u = \frac{4\sigma T^4}{c}$$

The quantity $a \equiv 4 \quad / c \approx 7.5657 \times 10^{-15} \, erg \, cm^{-3} K^{-4}$ is called the radiation constant. The frequency v_{max} at which B_v, the brightness per unit frequency of a blackbody, is maximum is the solution of,

$$\frac{\partial B_v}{\partial v} = 0;$$

It is,

$$\frac{v_{max}}{GHz} \approx 59\left(\frac{T}{K}\right)$$

Note that v_{max} is smaller than $c / \lambda_{max} \approx 103 GHz \, T(K)$,

where,

$$\left(\frac{\lambda_{max}}{cm}\right) \approx 0.29\left(\frac{K}{T}\right)$$

is the wavelength at which B, the brightness per unit wavelength, is highest. The latter equation is the familiar form of Wien's Law used by optical astronomers, whose spectrometers measure wavelengths instead of frequencies.

The Electrical Power Generated by a Warm Resistor

A resistor is any electrical device that absorbs all of the electrical power applied to it; it is the "black body" of electric circuits. Motions of charged particles in a warm resistor at temperature T>0 K generate electrical noise. The frequency spectrum of the noise power depends only on the resistor temperature and is independent of the resistor material. Also, the electrical noise generated is indistinguishable from the noise coming from an antenna observing a blackbody radiator. Warm resistors are useful in radio astronomy as standards for calibrating receiver gains, and the power received by a radio telescope is often described in terms of the "antenna temperature", the resistor temperature required to generate the same power thermally. The gain of a radio receiver can be calibrated by connecting it alternately to hot and cold resistors (called "loads") having known temperatures, and the amount of noise generated in a receiver can be described by the "receiver temperature", the temperature of a resistor at the input of an imaginary noiseless receiver having the same gain that would generate the same noise power output.

The derivation of the electrical power per unit bandwidth P_v generated by a resistor is a one-dimensional version of the three-dimensional argument used for blackbody radiation [Nyquist, H. 1928, Phys. Rev. 32, 110; Barrett, A. H. 1970, Fundamentals of Radio Astronomy, p. 22-23 (unpublished)]. At radio frequencies, it is often true that hv$\ll$kT and the Rayleigh-Jeans limit apply. Recall that the Rayleigh-Jeans derivation of B_v starts with a large cube of side length containing standing waves of thermal radiation. The average energy in each standing-wave mode is {E} =kT, and the number of modes with frequency v to v +d is proportional to v^2,$B_v \propto B^2$. For the power generated by a resistor, consider two identical resistors at temperature T connected by a lossless transmission line (e.g., a pair of parallel wires) of length a much larger than the longest wavelength of interest.

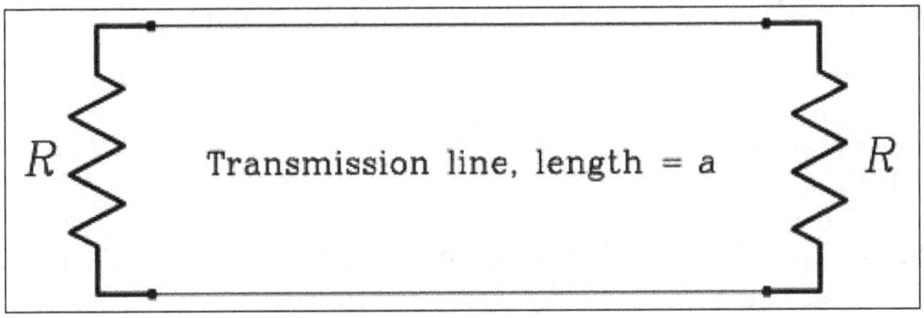

We assume that the transmission line has characteristic impedance equal to the resistance of the resistors so that power is efficiently coupled between them. In equilibrium, the transmission line can only support those standing waves having zero voltages at the ends; other modes are suppressed by the lossy resistors. The surviving standing waves satisfy,

$$a=\frac{n\lambda}{2}, \qquad n=1,2,3,...,$$

Where λ is the wavelength? Electrical signals do not travel at exactly the speed of light on a transmission line, but at some slightly lower velocity v<c. Thus,

$$n=\frac{2av}{v}$$

and, for a> λ , the number of modes per unit frequency is,

$$\frac{\Delta n}{\Delta v}=\frac{2a}{v}$$

In thermal equilibrium, the classical Boltzmann law says that each mode has average energy {E}=kT, so the average energy per unit frequency E_v in the transmission line is,

$$E=\frac{n}{v}kT=\frac{akT}{v}$$

This energy takes a time Δt=a/v to flow from one end of the transmission line to the other, so the classical power (energy per unit time) per unit frequency flowing on the transmission line is,

$$P_v=\frac{E_v}{\Delta t}=2kT$$

Note that the velocity P_v has dropped out. The total spectral power Pgenerated by the two identical resistors must be 2kT and, by symmetry, the spectral power generated by each resistor is,

$$P_\upsilon = kT$$

in the limit $h\upsilon \ll kT$ This equation is called the Nyquist formula and is the electrical equivalent of the Rayleigh-Jeans law for radiation. Since the "space" of the transmission line has only one dimension instead of three, the frequency dependence is v^o instead of v^2.

Still, the Nyquist formula suffers from an "ultraviolet catastrophe" of its own: the total power integrated over all frequencies diverges. Planck's quantization rule is the cure for this problem also: the electrical energy in each mode must be an integer multiple of hv, and the corrected version of the Nyquist formula is,

$$P_v = kT\left[\frac{hv / kT}{\exp[hv / (kT]-1}\right]$$

where the quantity in large brackets is the same quantization correction previously found for blackbody radiation. This equation is usually written in the form,

$$P_\upsilon = \frac{h\upsilon}{\exp\left(\frac{h\upsilon}{kT}\right)-1}$$

In many radio-astronomical applications, $h\upsilon \ll kT$ and the simpler P_v=kT (Eq. 2C6) can be used.

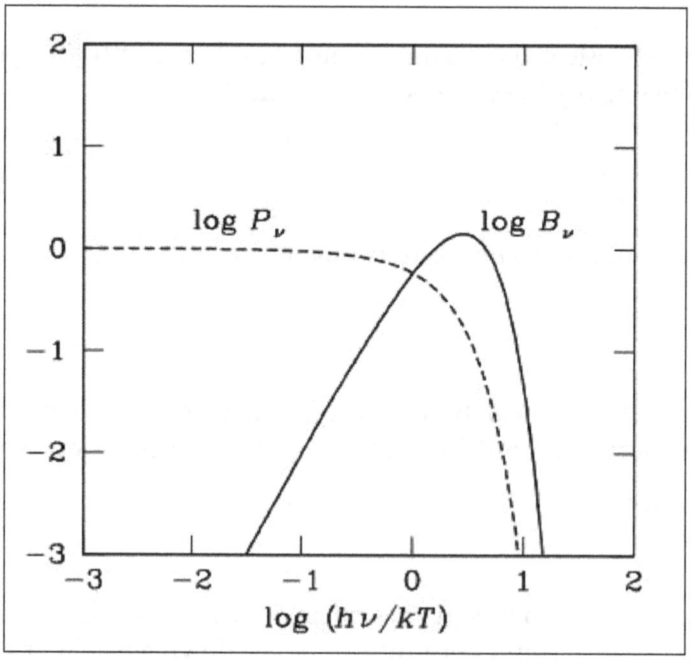

At low frequencies $\upsilon \ll kT / h$, the specific intensity B_v of blackbody radiation in three dimensions (solid curve) is proportional to v^2. Its one-dimensional analog, the spectral power density P_v of noise generated by a resistor (dashed curve), is proportional to v^0. Quantization causes both curves to cut off exponentially at high frequencies.

Luminosity

Luminosity refers to the total amount of energy produced by different celestial bodies(stars, galaxy) per unit time and it is basically measured in joules per second or watts in SI units. We can define luminosity as:

The absolute measure of radiated electromagnetic power (light), the radiant power emitted by a light-emitting object.

Values of luminosity are given in terms of the luminosity of the sun or in terms of magnitude which is called the absolute bolometric magnitude of an object is the measure of the total energy emission rate. A bolometer can be used to measure the radiant energy by the method of absorption and measurement of heating.

Factors

Luminosity generally depends upon two factors:

- The size of the star: The larger a star is, the more energy it puts out, and the more luminous it is.
- The temperature of the star: If two stars are of the same size but have different temperatures, then the star with a higher temperature will be more luminous than the star with lower temperatures.

The Brightness of a Star

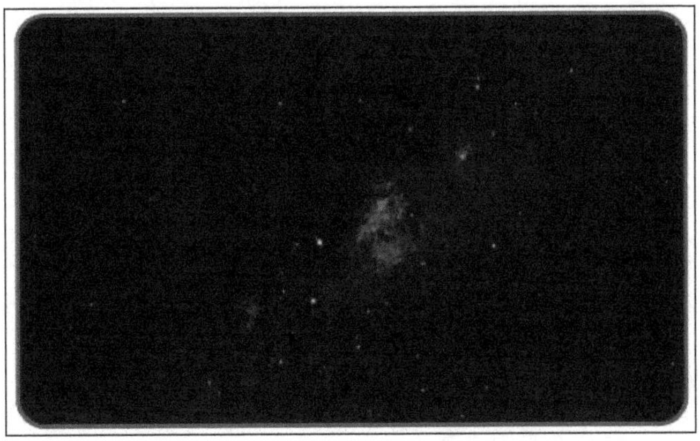

The brightness of a star which appears to our naked eyes mainly depends upon two factors:

- The luminosity of the star.
- The distance of the star from the Earth.

The Apparent Brightness of a source is a consequence of geometry. As light rays emerge from a source, they spread out in an area. According to the Inverse Square Law of brightness, the Apparent Brightness (B) of a source is inversely proportional to the square of its distance (d).

Mathematically,

$$B \propto \frac{1}{d^2}$$

Luminosity Theory

Luminosity depends on the surface area of the star. If the radius of a star is R then,

The Surface Area of the Star = 4PR2

Two stars having the same temperature, one with radius 2R will have 4 times greater luminosity than a star with radius R. The luminosity of a star also depends upon its temperature. When considering a star to be a completely black body, the radiation emitted per second will be according to the Stefan- Boltzmann law.

Energy Emitted Per Second (E) = sAT4

Where,

- s= Stefan's constant with a value of $5.7\times10^{-8}Wm^{-2}K^{-4}$.
- A= Surface Area of the Star.
- T = absolute temperature of the star.

Calculating the energy output for a star that is of the same size as the sun.

- $R = 6.96\times10^{8}$ m.
- T = 6000 K.

So, surface area of sun = surface area of star = $A = 4\pi r^2$

$=4\times\pi\times(6.96\times10^{8})$

$=6.08\times10^{18}$

Therefore, $E=5.7\times10^{-8}\times6.08\times10^{18}\times6000^{4}=4.5\times10^{26}J$.

A star having twice the temperature of Sun would have 16 times greater energy output.

Example of Luminosity

The Sun has a surface temperature of 6000K produces radiation with lmax = 420 nm. Find out the temperature of the Sirius if l max of Sirius is 72 nm.

Solution:

$\lambda_{max1}T_1=\lambda_{max2}T_2$

Therefore,

$420\times10^{-9}\times6000=72\times10^{-9}\times T_2$

$T_2=35000K$

Flux

The word "flux" in the context of stellar atmosphere theory differs from the "flux" of standard terminology. The symbol used is F, occasionally but not always printed in a special font. Let us imagine a horizontal surface embedded somewhere in a stellar atmosphere. It is being irradiated from below and above, but rather more from below than from above. Let us concentrate our attention for the time being on the radiation that is coming up from below. The rate of arrival of radiant energy per unit area from below from all directions would, in "standard" nomenclature, have been called the irradiance. After passage through the surface, it would be called the exitance. Now, using the nomenclature of stellar atmosphere theory, we call the rate of upward passage of radiant energy per unit area through a horizontal surface within the atmosphere the upward or the outward flux. The symbol is F+ (sometimes printed in a special font), and the SI units would be W m^{-2}. Likewise, the rate of passage of radiant energy per unit area from above to below is the downward or inward flux, F_-. The net upward flux is F = F+ – F−. If you are standing upright, F_+ is the irradiance of the soles of your feet, while F_- is the irradiance of the top of your head. If we measure the spherical angle θ from a downwardly-directed z-axis, then, following equation, we have,

$$F_+ = \int_0^{2\pi} \int_0^{\pi/2} I(\theta,\phi)\cos\theta\sin\theta d\theta\phi$$

$$F_- = \int_0^{2\pi} \int_0^{\pi/2} I(\theta,\phi)\cos\theta\sin\theta d\theta\phi$$

$$F = F_+ - F_- = \int_0^{2\pi} \int_0^{\pi/2} I(\theta,\phi)\cos\theta\sin\theta d\theta\phi$$

which is sometimes written for short,

$$F = \int_{4\pi} I\cos\theta d\omega \quad or\ just \quad \int I\cos\theta d\omega$$

At the surface of the star (if there is such a thing!) F_- = 0, so that F = F_+ = πI. This is the same as equation. At the center of the star, I is isotropic and F = 0.

Magnitude

In Astronomy, Magnitude is the measure of the brightness of a star or other celestial body. The brighter the object, the lower the number assigned as a magnitude. In ancient times, stars were ranked in six magnitude classes, the first magnitude class containing the brightest stars. In 1850 the English astronomer Norman Robert Pogson proposed the system presently in use. One magnitude is defined as a ratio of brightness of 2.512 times; e.g., a star of magnitude 5.0 is 2.512 times as bright as one of magnitude 6.0. Thus, a difference of five magnitudes corresponds to a brightness ratio of 100 to 1. After standardization and assignment of the zero point, the brightest class was found to contain too great a range of luminosities, and negative magnitudes were introduced to spread the range.

Apparent magnitude is the brightness of an object as it appears to an observer on Earth. The Sun's apparent magnitude is −26.7 that of the full Moon is about −11, and that of the bright star Sirius, −1.5. The faintest objects visible through the Hubble Space Telescope are of (approximately) apparent magnitude 30. Absolute magnitude is the brightness an object would exhibit if viewed from a distance of 10 parsecs (32.6 light-years). The Sun's absolute magnitude is 4.8.

Bolometric magnitude is that measured by including a star's entire radiation, not just the portion visible as light. Monochromatic magnitude is that measured only in some very narrow segment of the spectrum. Narrow-band magnitudes are based on slightly wider segments of the spectrum and broad-band magnitudes on areas wider still. Visual magnitude may be called yellow magnitude because the eye is most sensitive to light of that colour.

2

Branches of Astrophysics

There are many branches of Astrophysics such as observational astrophysics, theoretical astrophysics, nuclear astrophysics, computational astrophysics, etc. This chapter deals with each branch and its specific subject matter in a clear and concise manner for the benefit of the reader.

Observational Astrophysics provides a comprehensive and accessible introduction to the whole of modern astrophysics beyond the Solar System. It combines a critical account of observational methods (telescopes and instrumentation) with a lucid description of the Universe, including stars, galaxies and cosmology.

Observational Astrophysics

Observational Techniques

How does one begin to measure the distance to the stars and galaxies? If the parallax angle of a star is known then its distance can be calculated by trigonometry. However, the parallax angles are very small and cannot be measured from the ground to accuracies of better than 0.01.

Other parallax methods involve the motion of the sun among nearby stars. The sun moves relative to the fixed stars at a velocity of 20 km s-1 or about 4.1 AU per year in the direction of the constellation Hercules. However, the stars do not remain stationary and are all moving with their angular displacements or proper motions across the sky just like the sun, so we have to look at a group of stars and calculate the mean parallax of the group. The way this is done is quite elaborate and involves intricate statistical calculations, but the result is that we can reliably measure distances to a few 100 pc using this statistical parallax method. The Hubble Space Telescope can make parallax measurements to a precision of 0.001" and the recently launched Hipparchus satellite to 0.002".

The Period Luminosity (P-L) Relation

When we observe stars we find that there are some which vary in luminosity with time. These stars are called variables. The star Delta Cephei is one such star and its variation in luminosity with time has been extensively studied. There are many other stars like Delta Cephei and they go under the collective name of Cepheid Variables or cepheids.

It turns out that the longer the period the more luminous the cepheid star. The P-L relation was calibrated from apparent to absolute magnitudes by measuring the distance to a single Cepheid by statistical parallax methods. We can use the P-L relation to find the distance to the cepheid using the following steps:

- Step 1: Locate a cepheid.
- Step 2: Measure its period by observing it nightly.
- Step 3: Find the star's absolute magnitude using the P-L relationship.
- Step 4: Measure the star's received light flux to determine its apparent magnitude.
- Step 5: Calculate the star's distance using the distance-magnitude relation.

Variable stars that are used in this way to estimate distances are an example of what are called standard candles; objects that can be used as 'yardsticks' to determine astronomical distance scales.

Measuring Astronomical Distances

Since in most cases, the radiation received from celestial objects is weak, collectors of radiation are needed so that enough energy can be gathered for measurements to be made. Most visual and radio astronomy can be done from the ground, and once the energy is collected it must be analysed and detected so that information can be extracted from it.

Optical Telescopes

The invention of the optical telescope revolutionised our knowledge of astronomy by making distant objects appear closer and therefore larger. There are many different designs of telescope but all use an objective either in the form of a lens or a reflecting mirror, to concentrate starlight. Three important features of a telescope are it's:

- Magnification.
- Light gathering power.
- Image resolution.

A telescope is rather like a bucket for collecting photons. Astronomers like big telescopes because of their ability to gather light, and a measure of how much light one can collect is called the Light Gathering Power (LGP). The LGP of a telescope is directly proportional to the square of the diameter of its objective and is a relative measure for comparing the ability of different telescopes to 'grasp' light.

Although the atmosphere readily admits electromagnetic radiation at optical wavelengths, in practice the 'seeing' is limited by atmospheric conditions particularly by convection currents and air turbulence as well as minute changes in the primary mirror shape due to structural flexing. Two techniques, active optics and adaptive optics have been devised to overcome these. In active optics, alterations in the mirror shape due to mechanical stresses are made in a few minutes by mechanical actuators to ensure that the image is kept in the sharpest focus.

Adaptive optics is concerned with compensating for the effects of atmospheric turbulence. The optical wave fronts from a reference star are distorted as they travel through the atmosphere. These distortions are measured and the information sent by a computer to piezoelectric transducers that distort the mirror of the telescope so that it matches that of the turbulent atmosphere. The effect is to cancel out the effect of the turbulence resulting in better image resolution.

Radio Telescopes

Radio astronomy was born when a telephone engineer Karl Jansky was looking for sources of static noise affecting radiotelephone communications. Using a radio aerial, Jansky found that some of this noise was coming from the Milky Way. Later, other workers found numerous celestial radio sources including the sun and radio telescopes were developed to study them.

The simplest radio telescope consists of a 'flux bucket' in which radio energy is concentrated in a parabolic dish aerial and brought to a focus at a receiver. A radio telescope has a low resolving power because of the long wavelengths of radio waves. Remember that the resolving power is proportional to the wavelength, so that for an optical telescope and a radio dish of the same aperture, the radio telescope will have at least 100,000 times less resolution.

However there is a way to increase the resolving power without having to build huge dishes. Radio astronomers use interferometry to improve the image resolution. If a celestial source is directly overhead then the signals from it arrive at the aerials in phase and constructively interfere giving a strong signal. Conversely, when the signals are 180° out of phase the signals destructively interfere. As the source moves across the sky an interference pattern of maxima and minima are recorded exactly like that for light passing through a double slit. In order to build up a complete image of the source we need to have interferometers at different orientations. The Very Large Array in New Mexico is able to map the sky at radio wavelengths. This technique of mapping a complete object is called aperture synthesis and involves sophisticated signal processing.

Infrared Astronomy

Observing in the infrared is limited to only a few wavelength ranges due to absorption by CO_2 and water vapour in the Earth's atmosphere. An optical telescope is used with a special detector called a Bolometer placed at its focus. The telescope concentrates the infrared object onto the bolometer from which its temperature can be determined. Since the infrared observations are restricted from the ground, much infrared astronomy is done using space astronomy, the IRAS satellite being a notable example.

Ground Based Observing Techniques

The use of high altitude balloons, rockets, satellites and spacecraft enables us to observe space at wavelengths that would not normally be absorbed by the Earth's atmosphere. This method of observation usually goes under the name of Space Astronomy.

X-Ray Astronomy

A rocket borne experiment launched in 1962 provided the first direct evidence that some celestial objects are strong X-ray emitters. Detectors aboard the rocket revealed a sharp increase in X-ray photon flux as they were pointed in the direction of Scorpius and also in the constellations of Sagittarius and Cygnus. In addition the photon flux never dropped to zero indicating that the sky was immersed in a background of interstellar origin. Subsequent experiments flown on spacecraft have revealed more information about the number and disposition of X-ray sources in the universe and located many more X-ray sources. X-ray sources are of great interest to astrophysicists as they are associated with highly energetic processes involving very high temperatures. Nearly all types of astronomical objects from planets to stars and galaxies are found to emit X-rays.

Gamma Ray Astronomy

During the cold war, the US military launched a series of satellites called Vela that were designed to ensure compliance with the 1963 nuclear test ban treaty. Vela were able to detect the flash of gamma rays given off by the detonation of a nuclear weapon. The Vela satellites did indeed detect gamma ray flashes but they were not coming from Earth but from space! These gamma ray bursters (GBRs) last from about 0.01 to 1000 s with energies ranging between 1 KeV to 100 MeV. In 1991 the space shuttle Atlantis released the Compton Gamma Ray Observatory (CGRO). This satellite is equipped with sensitive gamma ray detectors that can assign position information in order to correlate the bursts with known objects. The CGRO has found that the sources are spread evenly across space but their distribution is not uniform. The origin of GRB's is currently a mystery and it is not known whether they originate within the galaxy or outside it.

Space Based Observing Techniques

Detecting and recording electromagnetic radiation from space has far advanced from the time when drawings were made by astronomers looking through the eyepiece of a telescope. Modern technology has extended the range of wavelengths and intensities normally accessible to the human eye and the processing of images by computers can reveal information and structure not immediately discernible to the brain.

Photographic Emulsion

Photography is extensively employed to record astronomical images. A photographic emulsion is made up of lots of tiny grains a few microns across. The chemical development of these grains after being activated by incident photons is what forms the image and the grain size determines the image resolution. The emulsion is coated onto a glass plate which does not flex or expand easily and from which positions and intensities of stars can be accurately measured. Photographic plates have long storage lifetimes and emulsions can also be optimised for certain wavelengths, with the resulting image digitised for input into a computer.

An important measure of detectors sensitivity to electromagnetic radiation is its Quantum Efficiency QE. This is defined as:

$$QE = \frac{\text{Number of photons detected}}{\text{Number of photons incident}} \times 100\%$$

An ideal detector would have a QE of 100% and the QE of photographic emulsion is quite low being only a few percent. However it is relatively cheap, simple to use and it can compensate for its low Q if exposures are made over long time periods or integrating the light received. It also presents a large sensitive surface area to the image.

Photomultiplier Tubes (PMTs)

These make use of the photoelectric effect. An incident photon striking the surface of certain metals will eject electrons from it if it has a certain minimum energy given by the Planck relation.

$$E = hf - W$$

where E and f are the respective energy and frequency of the incident photon, and W is the work function which is the energy needed to remove the electron from the surface of the metal. A photomultiplier consists of a series of dynodes arranged in an evacuated glass tube. Light falling on the first dynode ejects electrons by the photoelectric effects which are then accelerated through a potential difference to the next dynode where more electrons are released by secondary emission. As a result an electron amplification process occurs at each diode stage right through to the final electrode where a useful current is produced. Photomultipliers do not produce images but measure intensities with a QE value as high as 20%.

Gamma Ray Detection

PMTs are used in conjunction with sodium iodide crystals to detect gamma rays. When struck by a gamma ray the sodium iodide crystal emits a flash of light which is detected by the PMT and amplified. The gamma ray detectors on CGRO incorporate sodium iodide crystals and PMT's as a means of detecting gamma ray bursts.

Charge-Coupled Devices

A charge-coupled device (CCD) is a type of microchip in which incident light is converted directly into digital information. CCDs are made out of a silicon wafer divided into small regions called pixels. A typical CCD may have as many as a million pixels extending over an area of a few cm2 arranged in rows and columns. When light strikes the CCD, electric charge is accumulated in pixels which are proportional to the brightness of the image at a particular pixel location. Unlike photographic film, this makes the response of a CCD linear. The QE of a CCD is close to 100% making it a near ideal radiation detector. This means that the light gathering power of a smaller telescope equipped with a CCD as a detector, gives comparable performance with a much larger one using photographic film.

The fact, the CCDs output their information digitally makes them ideally suited to computer processing and data transmission. The images you see taken by the HST are recorded using CCDs. The WFPC and FOC employ 800 x 800 pixel CCD cameras, and the readouts are transmitted to Earth in digital form for image processing.

Bolometers

A Bolometer is a device which measures increases in temperature due to the radiant energy it receives at all wavelengths. There are two types; thermistors which operate at room temperature, and semiconductor bolometers made of germanium, which need to be cooled to low temperatures ~2K. In both kinds, as the device heats up, its electrical resistivity is altered and its resistance to an electric current also changes. The variation of current with temperature indicates how much energy the bolometer is absorbing. Bolometers are extensively used in infrared astronomy. The IRAS telescope uses infrared bolometric detectors cooled by liquid helium to 2K which operated for 9 months before the helium ran out.

Detectors and Imaging Devices

In general there are four types of analyzers used for astrophysical measurements:

- Spectrometers for measuring the spectra of starlight.
- Photometers for measuring the brightness of stars.
- Polari meters for measuring the degree of polarization of starlight.
- Interferometers for high precision angular measurements of the separation of double stars and the diameter of single ones.

Spectrometers

Spectrometers are instruments that split light into its component wavelengths. This is accomplished by using prisms or diffraction gratings to disperse light into a spectrum. For light of a particular wavelength h, the performances of a spectrometer are governed by two criteria:

$$\text{The dispersion} = \frac{\text{the change in wavelength}}{\text{the change in distance along the spectrum}}$$

$$\text{and the spectral resolution R} = \frac{\text{the wavelength}}{\text{smallest wavelength interval that can be measured on the film}}$$

Higher dispersions increase the amount of detail we can see in the spectrum but also dilute the light and we can only obtain high dispersion spectra for the brighter stars using telescopes with large apertures. Typical dispersions for astronomical work range from 200 nm mm-1 and spectral resolutions of 10 to 10 000. For prism spectrometers, relatively low levels of starlight must travel through a thick prism causing much light to be absorbed. In addition, the spectrum of light is not dispersed evenly- the red end compressed while the blue and violet is spread out.

The spectrometer more commonly used for astronomical work is the grating spectrometer. This uses a diffraction grating as the dispersion element. A disadvantage of grating spectrometers is that the light energy is spread over several spectra corresponding to different orders rather than concentrated into single spectra by a prism spectrometer. However, the resolving power of gratings are generally better than for a prism and by using reflection gratings it is possible to analyse wavelengths that would normally be absorbed by a glass prism.

Photometers

Measuring the light intensity or magnitude of stars with detectors is called photometry and an instrument which does this is called a photometer. The eye, photographic film, CCDs and bolometers can all be used for photometric measurements. Photomultiplier tubes are commonly employed as photometers in conjunction with an arrangement of filters called the UBV filter system.

The bolometric magnitude is the apparent magnitude of a star measured above the Earth's atmosphere at all wavelengths. However in practice, astronomers measure the radiant energy of a star through three filters U, B and V transparent to a specific wavelength. These are defined as:

- U, being a filter transparent to a star's Ultraviolet apparent magnitude centered at a wavelength of 356 nm with a bandwidth either side of 68 nm.
- B, being the star's Blue apparent magnitude as seen through a filter centered on 440 nm with a bandwidth of 98 nm.
- V, being the star's apparent Visual magnitude as seen through a filter centered on 550 nm with a bandwidth of 89 nm. The V filter approximates the sensitivity of the human eye.

Astronomers use photometry to determine the color of a star so that they can estimate its temperature and find out how hot it is. The temperature of a star is determined by aiming a telescope at it and measuring the intensity of the light as it passes through each of the filters in turn. The relative intensities in neighboring wavelengths are compared by subtracting the ultraviolet and blue magnitudes (U-B) and the blue and visual magnitudes (B-V. These are called the stars color indices. We can therefore write:

$$B\text{-}V=M_B\text{-}M_V$$

and

$$U\text{-}V=M_U\text{-}M_V$$

Note that because a color index is the difference of two magnitudes, we vdo not need to know the star's distance. The color index of a star tells you how brighter or dimmer a star is in one wavelength band than in another. Assuming that stars are near ideal blackbodies, we know from Wien's law that the hotter a star is, the more its peak wavelength is skewed towards the ultraviolet, so that the magnitude is greater through the U filter than through the B and V filters. If the star is relatively cool then its magnitude will be greater in the V than in the B and U filters. From the laws of blackbody radiation, it can be shown that a mathematical relationship exists between the temperature of a star and its color index.

Photographic Methods

The images of stars on a photographic plate can be measured photo metrically using an instrument called a microdensitometer. This consists of a device which shines a beam of light through the image and measures the transmitted intensity as a variation in voltage. Since the light response of photographic emulsion is non-linear the response curve must be known beforehand so that the conversion of the stellar image into intensity is properly calibrated.

If we have a microdensitometer which transmits light of intensity Iin through an image in which transmitted intensity is I_{out} then these are related to the to the photographic intensity D by,

$$D=\log I_{in} - \log I_{out}$$

and using the characteristic curve we can relate D to the original image brightness.

Polarimeters

Light from distant stars is polarised as it passes through clouds of interstellar gas and dust in the Milky Way. Dust grains are aligned by the presence of magnetic fields which causes the polarisation of starlight as it passes through. The degree of polarisation is measured with a polarimeter and from this we can deduce the magnetic field strength existing inside the clouds.

Interferometers

A Michelson Stellar Interferometer is capable of measuring the angular separations of stars and even the angular size of the nearest stars. The stellar interferometer works in the same way as the Young's double slit experiment. In Young's experiment, light passes through two slits separated by a distance d. In a stellar interferometer, the light sources can be two stars, or light from opposite ends of a star along its diameter. Unlike the Young's slit experiment however, the light from the two sources is not coherent. The two independent fringe sets simply cancel out due to the bright and dark fringes overlapping.

Image Processing

In the last 20 years, there has been a revolution in computing power with computers increasing in processing ability while at the same time decreasing in cost. This, together with technological advances in acquiring astronomical data in digital form using detectors such as CCDs, has meant that the manipulation of images inside a computer has become routine tool in extracting the maximum amount of information from an astronomical object. Some common image processing techniques are:

- Noise removal: Noise may be digitally removed and replaced with the average intensity of their adjacent areas.
- Background subtraction: If the extent of the background signal is known, then this is a systematic error that can be subtracted out in the final image.

- Contrast Stretch: A typical computer can display 16 levels of grey while a CCD can resolve more than 256 changes in intensity. In order to make use of the CCD's greater range of levels, a contrast search is applied. This involves mapping the dominant grey levels in the image to the maximum range of the computer display.
- False color: Sometimes it is helpful to assign colours from the computers colour palette to particular intensities on the image. This is particularly useful when we want to find for example, the temperature distribution of an object or the intensity variation of radiation across the sky.

Analysing Radiation from Space

Theoretical astrophysics is a curious discipline, more like a classification than a unified field of study. A "theoretical astrophysicist" might study the dynamics of planetary rings, or the microphysics of interstellar dust, or the formation of stars, or the evolution of the universe.

Theoretical Astrophysics

There is no way of distinguishing between the effects on an observer of a uniform gravitational field and of constant acceleration. The Principle of the Equivalence of Gravitation and Inertia provides rests on the equality of gravitational and inertial mass, demonstrated by Galileo, Huygens, Newton, Bessel, and Eotvos. Einstein concluded that one could not detect the difference between a uniform static gravitational field and uniform acceleration. Einstein elevated this concept to become the Principle of Equivalence which is the foundation of the General Theory of Relativity.

The equivalence principle can be stated as "At every space-time point in an arbitrary gravitational field, it is possible to choose a locally inertial coordinate system such that, within a sufficiently small region of the point in question, the laws of nature take the same form as in unaccelerated Cartesian coordinate systems.

The Weak Equivalence Principle

A key stone to Einstein's theory of General Relativity Theory is the principle of the uniqueness of free fall.

The Principle of the Uniqueness of Free Fall

The world line of a freely falling test body is independent of its composition or structure. The uniqueness of free fall trajectories allows one to regard space-time as filled as filled with a set of curves, the test body trajectories, which are unique aside from parameterization. When translated into Newtonian language, the uniqueness of free fall states than any two test bodies must fall with the same acceleration in a given external gravitational field.

A short form of the two principles of equivalence is:

- The Strong Principle of Equivalence states all the laws of nature are the same in a uniform static gravitational field and the equivalent accelerated reference frame.

- The Weak Principle of Equivalence states all the laws of motion for freely falling particles are the same as in an unaccelerated reference frame.

That is the Weak Equivalence Principle is a restatement of the equality of gravitational and inertial mass.

Neutrino Test of the Equivalence Principle

If a Gamma Ray Burster occurs at a distance of about 100 MPc (326 million light years), observation of both the time of arrival of the gamma rays (via a GRB monitoring satellite such as currently operating GRO) and detection of ultra-high energy neutrinos (e.g PeV energy) would allow a test of the Weak Equivalence principle to about the 10^ {-11} level on cosmic scale in a direct test of the uniqueness of free fall.

Thermodynamics of Black Holes

During the past 30 years, research in the theory of black holes in general relativity has brought to light strong hints of a very deep and fundamental relationship between gravitation, thermodynamics, and quantum theory. The cornerstone of this relationship is black hole thermodynamics, where it appears that certain laws of black hole mechanics are, in fact, simply the ordinary laws of thermodynamics applied to a system containing a black hole. Indeed, the discovery of the thermodynamic behavior of black holes — achieved primarily by classical and semiclassical analyses — has given rise to most of our present physical insights into the nature of quantum phenomena occurring in strong gravitational fields.

Review of the following aspects of black hole thermodynamics:

- At the purely classical level, black holes in general relativity (as well as in other diffeomorphism covariant theories of gravity) obey certain laws which bear a remarkable mathematical resemblance to the ordinary laws of thermodynamics. The derivation of these laws of classical black hole mechanics is reviewed in section.

- Classically, black holes are perfect absorbers but do not emit anything; their physical temperature is absolute zero. However, in quantum theory black holes emit Hawking radiation with a perfect thermal spectrum. This allows a consistent interpretation of the laws of black hole mechanics as physically corresponding to the ordinary laws of thermodynamics. The status of the derivation of Hawking radiation is reviewed in section.

- The generalized second law (GSL) directly links the laws of black hole mechanics to the ordinary laws of thermodynamics. The arguments in favor of the GSL are reviewed in section. A discussion of entropy bounds is also included in this section.

- The classical laws of black hole mechanics together with the formula for the temperature of Hawking radiation allow one to identify a quantity associated with black holes — namely A/4 in general relativity — as playing the mathematical role of entropy. The apparent validity of the GSL provides strong evidence that this quantity truly is the physical entropy of a black hole. A major goal of research in quantum gravity is to provide an explanation for — and direct derivation of — the formula for the entropy of a black hole. A brief survey of work along these lines is provided in section.

- Although much progress has been made in our understanding of black hole thermodynamics, many important issues remain unresolved. Primary among these are the "black hole information paradox" and issues related to the degrees of freedom responsible for the entropy of a black hole. These unresolved issues are briefl discussed in section.

Throughout this article, we shall set G = ħ = c = k = 1, and we shall follow the sign and notational conventions of. Although I have attempted to make this review be reasonably comprehensive and balanced, it should be understood that my choices of topics and emphasis naturally reflect my own.

Classical Black Hole Thermodynamics

In physical terms, a black hole is a region where gravity is so strong that nothing can escape. In order to make this notion precise, one must have in mind a region of space time to which one can contemplate escaping. For an asymptotically flat space time $(M; g_{ab})$ (representing an isolated system), the asymptotic portion of the space time "near infinity" is such a region. The black hole region, B, of an asymptotically flat space time, $(M; g_{ab})$, is defined as:

$$B \equiv M - I^{-}(I^{+})$$

where I^{+} denotes future null infinity and I^{-} denotes the chronological past. Similar definitions of a black hole can be given in other contexts (such as asymptotically anti-deSitter spacetimes) where there is a well-defined asymptotic region.

The *event horizon*, H, of a black hole is defined to be the boundary of B. Thus, H is the boundary of the past of I^{+}. Consequently, H automatically satisfies all of the properties possessed by past boundaries. In particular, H is a null hyper surface which is composed of future in extendible null geodesics without caustics, i.e., the expansion, θ, of the null geodesics comprising the horizon cannot become negatively infinite. Note that the entire future history of the space time must be known before the location of H can be determined, i.e., H possesses no distinguished local significance.

If Einstein's equation holds with matter satisfying the null energy condition (i.e., if $T_{ab}k^{a}k^{b} \geq 0$ for all null k^{a}), then it follows immediately from the Raychauduri equation that if the expansion, θ, of any null geodesic congruence ever became negative, then θ would become infinite within a finite affine parameter, provided, of course, that the geodesic can be extended that far. If the black hole is *strongly asymptotically predictable* — i.e., if there is a globally hyperbolic region containing $I^{-}(I^{+}) \rightarrow$ H — it can be shown that this implies that $\theta \geq 0$ everywhere on H It then follows that the surface area, *A*, of the event horizon of a black hole can never decrease with time, as discovered by Hawking.

It is worth remarking that since H is a past boundary, it automatically must be a C^{0} embedded submanifold, but it need not be C^{1}. However, essentially all discussions and analyses of black hole event horizons implicitly assume C^{1} or higher order differentiability of H. Recently, this higher order differentiability assumption has been eliminated for the proof of the area theorem.

The area increase law bears a resemblance to the second law of thermodynamics in that both laws assert that a certain quantity has the property of never decreasing with time. It might seem that this resemblance is a very superficial one, since the area law is a theorem in differential geometry whereas the second law of thermodynamics is understood to have a statistical origin. Nevertheless, this resemblance together with the idea that information is irretrievably lost when a body falls into a black hole led Bekenstein to propose that a suitable multiple of the area of the event horizon of a black hole should be interpreted as its entropy, and that a *generalized second law* (GSL) should hold: The sum of the ordinary entropy of matter outside of a black hole plus a suitable multiple of the area of a black hole never decreases.

The remaining laws of thermodynamics deal with equilibrium and quasi-equilibrium processes. At nearly the same time as Bekenstein proposed a relationship between the area theorem and the second law of thermodynamics, Bardeen, Carter, and Hawking provided a general proof of certain laws of "black hole mechanics" which are direct mathematical analogs of the zeroth and first laws of thermodynamics. These laws of black hole mechanics apply to stationary black holes.

In order to discuss the zeroth and first laws of black hole mechanics, we must introduce the notions of stationary, static, and axisymmetric black holes as well as the notion of a Killing horizon. If an asymptotically flat spacetime $(M; g_{ab})$ contains a black hole, B, then B is said to be *stationary* if there exists a one-parameter group of isometries on $(M; g_{ab})$ generated by a Killing field t^awhich is unit timelike at infinity. The black hole is said to be *static*if it is stationary and if, in addition, t^a is hypersurface orthogonal. The black hole is said to be *axisymmetric* if there exists a one parameter group of isometries which correspond to rotations at infinity. A stationary, axisymmetric black hole is said to possess the "t-φ orthogonality property" if the 2-planes spanned by t^a and the rotational Killing field φ^a are orthogonal to a family of 2-dimensional surfaces. The t-φ orthogonality property holds for all stationary-axisymmetric black hole solutions to the vacuum Einstein or Einstein-Maxwell equations.

A null surface, K, whose null generators coincide with the orbits of a one-parameter group of isometries (so that there is a Killing field ξ^a normal to K), is called a *Killing horizon*. There are two independent results (usually referred to as "rigidity theorems") that show that in a wide variety of cases of interest, the event horizon, H, of a stationary black hole must be a Killing horizon. The first, due to Carter, states that for a static black hole, the static Killing field t^a must be normal to the horizon, whereas for a stationary-axisymmetric black hole with the t-φ orthogonality property there exists a Killing field ξ^a of the form.

$$\xi^a = t^a + \Omega\phi^a$$

which is normal to the event horizon. The constant Ω defined by Eq. ($\xi^a = t^a + \Omega\phi^a$) is called the *angular velocity of the horizon*. Carter's result does not rely on any field equations, but leaves open the possibility that there could exist stationary black holes without the above symmetries whose event horizons are not Killing horizons. The second result, due to Hawking, directly proves that in vacuum or electrovac general relativity, the event horizon of any stationary black hole must be a Killing horizon. Consequently, if t^a fails to be normal to the horizon, then there must exist an additional Killing field, ξ^a, which is normal to the horizon, i.e., a stationary black hole must be nonrotating (from which staticity follows or axisymmetric (though not necessarily with the t-φ orthogonality property). Note that Hawking's theorem makes no assumptions of symmetries beyond stationarity, but it does rely on the properties of the field equations of general relativity.

Now, let K be any Killing horizon (not necessarily required to be the event horizon, H, of a black hole), with normal Killing field ξ^a. Since $\nabla^a(\xi^b\xi_b)$ also is normal to K, these vectors must be proportional at every point on K. Hence, there exists a function, κ, on K, known as the *surface gravity* of K, which is defined by the equation.

$$\nabla^a\left(\xi^b\xi_b\right)\xi^a = -2k\xi^a$$

It follows immediately that κ must be constant along each null geodesic generator of K, but, in general, κ can vary from generator to generator. It is not difficult to show that,

$$k = \lim(\mathrm{Va})$$

where a is the magnitude of the acceleration of the orbits of ξ^a in the region off of K where they are timelike, $V \equiv (-\xi^a\xi_a)^{1/2}$ is the "redshift factor" of ξ^a, and the limit as one approaches K is taken. Equation ($4k = \lim(\mathrm{Va})$) motivates the terminology "surface gravity". Note that the surface gravity of a black hole is defined only when it is "in equilibrium", i.e., stationary, so that its event horizon is a Killing horizon. There is no notion of the surface gravity of a general, non-stationary black hole; although the definition of surface gravity can be extended to isolated horizons.

In parallel with the two independent "rigidity theorems", there are two independent versions of the zeroth law of black hole mechanics. The first, due to Carter, states that for any black hole which is static or is stationary-axisymmetric with the t-φ orthogonality property, the surface gravity κ, must be constant over its event horizon H. This result is purely geometrical, i.e., it involves no use of any field equations. The second, due to Bardeen, Carter, and Hawking states that if Einstein's equation holds with the matter stress-energy tensor satisfying the dominant energy condition, then κ must be constant on any Killing horizon. Thus, in the second version of the zeroth law, the hypothesis that the t-φ orthogonality property holds is eliminated, but use is made of the field equations of general relativity.

A *bifurcate Killing horizon* is a pair of null surfaces, K_A and K_B, which intersect on a spacelike 2-surface, C (called the "bifurcation surface"), such that K_A and K_B are each Killing horizons with respect to the same Killing field ξ^a. It follows that ξ^a must vanish on C; conversely, if a Killing field, ξ^a, vanishes on a two-dimensional spacelike surface, C, then C will be the bifurcation surface of a bifurcate Killing horizon associated with ξ^a. An important consequence of the zeroth law is that if $\kappa \neq 0$, then in the "maximally extended" spacetime representing a stationary black hole, the event horizon, H, comprises a branch of a bifurcate Killing horizon. This result is purely geometrical — involving no use of any field equations. As a consequence, the study of stationary black holes which satisfy the zeroth law divides into two cases: "extremal" black holes (for which, by definition, $\kappa = 0$), and black holes with bifurcate horizons.

The first law of black hole mechanics is simply an identity relating the changes in mass, M, angular momentum, J, and horizon area, A, of a stationary black hole when it is perturbed. To first order, the variations of these quantities in the vacuum case always satisfy.

$$\delta M = \frac{1}{8\pi}k\delta A + \Omega\delta J$$

In the original derivation of this law, it was required that the perturbation be stationary. Furthermore, the original derivation made use of the detailed form of Einstein's equation. Subsequently, the derivation has been generalized to hold for non-stationary perturbations, provided that the change in area is evaluated at the bifurcation surface, C, of the unperturbed black hole (see, however, for a derivation of the first law for non-stationary perturbations that does not require evaluation at the bifurcation surface). More significantly, it has been shown that the validity of this law depends only on very general properties of the field equations. Specifically, a version of this law holds for any field equations derived from a diffeomorphism covariant Lagrangian, *L*. Such a Lagrangian can always be written in the form.

$$L = L\left(g_{ab};R_{abcd},\nabla_a R_{bcde},\ \ldots;\psi,\nabla_a\psi,\ \ldots\right),$$

where ∇_a denotes the derivative operator associated with g_{ab}, R_{abcd} denotes the Riemann curvature tensor of g_{ab}, and ψ denotes the collection of all matter fields of the theory (with indices suppressed). An arbitrary (but finite) number of derivatives of R_{abcd} and ψ are permitted to appear in *L*. In this more general context, the first law of black hole mechanics is seen to be a direct consequence of an identity holding for the variation of the Noether current. The general form of the first law takes the form,

$$\delta M = \frac{k}{2\pi}\delta S_{bh} + \Omega\delta J + \ \ldots,$$

where the "..." denote possible additional contributions from long range matter fields, and where,

$$S_{bh} \equiv -2\pi\int_c \frac{\delta L}{\delta R_{abcd}} n_{ab} n_{cd}$$

Here n_{ab} is the binormal to the bifurcation surface C (normalized so that $n_{ab}n^{ab} = -2$), and the functional derivative is taken by formally viewing the Riemann tensor as a field which is independent of the metric in Eq. ($L = L\left(g_{ab};R_{abcd},\nabla_a R_{bcde},\ \ldots;\psi,\nabla_a\psi,\ \ldots\right),$). For the case of vacuum general relativity, where $L = R\sqrt{-g}$, a simple calculation yields,

$$S_{bh} = A/4$$

and Eq. ($\delta M = \frac{k}{2\pi}\delta S_{bh} + \Omega\delta J + \ \ldots,$) reduces to Eq. ($\delta M = \frac{1}{8\pi}k\delta A + \Omega\delta J$).

The close mathematical analogy of the zeroth, first, and second laws of thermodynamics to corresponding laws of classical black hole mechanics is broken by the Planck-Nernst form of the third law of thermodynamics, which states that $S \to 0$ (or a "universal constant") as $T \to 0$. The analog of this law fails in black hole mechanics — although analogs of alternative formulations of the third law do appear to hold for black holes — since there exist extremal black holes (i.e., black holes with $\kappa = 0$) with finite *A*. However, there is good reason to believe that the "Planck-Nernst theorem" should not be viewed as a fundamental law of thermodynamics but rather as a property of the density of states near the ground state in the thermodynamic limit, which happens to be valid for commonly studied materials. Indeed, examples can be given of ordinary quantum systems that violate the Planck-Nernst form of the third law in a manner very similar to the violations of the analog of this law that occur for black holes.

The zeroth and first laws of black hole mechanics have been formulated in the mathematical setting of stationary black holes whose event horizons are Killing horizons. The requirement of stationarity applies to the entire space time and, indeed, for the first law, stationarity of the entire space time is essential in order to relate variations of quantities defined at the horizon (like A) to variations of quantities defined at infinity (like M and J). However, it would seem reasonable to expect that the equilibrium thermodynamic behavior of a black hole would require only a form of local stationarity at the event horizon. For the formulation of the first law of black hole mechanics, one would also then need local definitions of quantities like M and J at the horizon. Such an approach toward the formulation of the laws of black hole mechanics has recently been taken via the notion of an isolated horizon, defined as a null hyper surface with vanishing shear and expansion satisfying the additional properties stated in. (This definition supersedes the more restrictive definitions given.

The presence of an isolated horizon does not require the entire space time to be stationary A direct analog of the zeroth law for stationary event horizons can be shown to hold for isolated horizons. In the Einstein-Maxwell case, one can demand (via a choice of scaling of the normal to the isolated horizon as well as a choice of gauge for the Maxwell field) that the surface gravity and electrostatic potential of the isolated horizon be functions of only its area and charge. The requirement that time evolution be symplectic then leads to a version of the first law of black hole mechanics as well as a (in general, non-unique) local notion of the energy of the isolated horizon. These results also have been generalized to allow dilation couplings and Yang-Mills fields.

In comparing the laws of black hole mechanics in classical general relativity with the laws of thermodynamics, it should first be noted that the black hole uniqueness theorems establish that stationary black holes – i.e., black holes "in equilibrium" – are characterized by a small number of parameters, analogous to the "state parameters" of ordinary thermodynamics. In the corresponding laws, the role of energy, E, is played by the mass, M, of the black hole; the role of temperature, T, is played by a constant times the surface gravity, κ of the black hole; and the role of entropy, S, is played by a constant times the area, A, of the black hole. The fact that E and M represent the same physical quantity provides a strong hint that the mathematical analogy between the laws of black hole mechanics and the laws of thermodynamics might be of physical significance. However, as argued in, this cannot be the case in classical general relativity. The physical temperature of a black hole is absolute zero, so there can be no physical relationship between T and κ. Consequently, it also would be inconsistent to assume a physical relationship between S and A.

Black Hole Entropy

The Bekenstein-Hawking entropy or black hole entropy is the amount of entropy that must be assigned to a black hole in order for it to comply with the laws of thermodynamics as they are interpreted by observers external to that black hole. This is particularly true for the first and second laws. Black hole entropy is a concept with geometric root but with many physical consequences. It ties together notions from gravitation, thermodynamics and quantum theory, and is thus regarded as a window into the as yet mostly hidden world of quantum gravity.

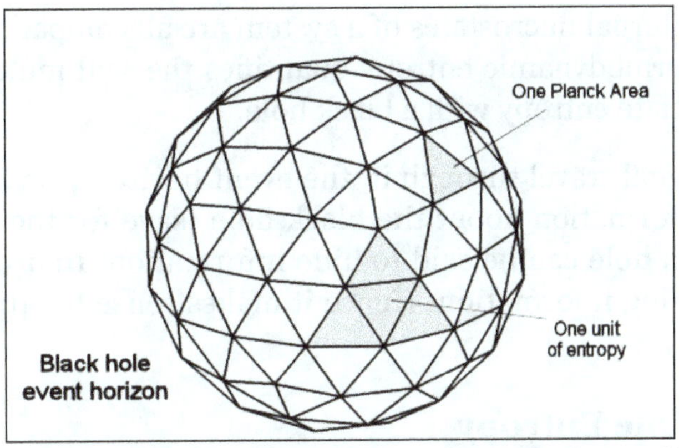

The Bekenstein-Hawking entropy is the entropy to be ascribed to any black hole: one quarter of its horizon area expressed in units of the Planck area.

Why Black Hole Entropy?

Due to the disposition of the local light cone, the event horizon stops any signals bearing interior information from exiting the black hole. Only the hole's mass M, angular momentum J and electric charge Q are sensed by an exterior observer.

A black hole may be described as a blemish in space time, or a locale of very high curvature. Is it meaningful or desirable to associate entropy with it? Is this possible at all? There are several ways to justify the concept of black hole entropy:

- A black hole is usually formed from the collapse of a quantity of matter or radiation, both of which carry entropy. However, the hole's interior and contents are veiled to an exterior observer. Thus a thermodynamic description of the collapse from that observer's viewpoint cannot be based on the entropy of that matter or radiation because these are unobservable. Associating entropy with the black hole provides a handle on the thermodynamics.
- A stationary black hole is parameterized by just a few numbers: its mass, electric charge and angular momentum (and magnetic monopole charge, except its actual existence in nature has not been demonstrated yet). For any specific choice of these parameters one can imagine many scenarios for the black hole's formation. Thus there are many possible internal states corresponding to that black hole. In thermodynamics one meets a similar

situation: many internal microstates of a system are all compatible with the one observed (macro) state. Thermodynamic entropy quantifies the said multiplicity. Thus by analogy one needs to associate entropy with a black hole.

- By blocking all signal travel through it, the event horizon prevents an external observer from receiving information about the black hole (save for the mentioned few parameters). Thus a black hole can be said to hide information. In ordinary physics entropy is a measure of missing information. Hence it makes sense to attribute entropy to a black hole.

Formula for Black Hole Entropy

How to express the black hole entropy in a concrete formula? It is clear at the outset that black hole entropy should only depend on the observable properties of the black hole: mass, electric charge and angular momentum. It turns out that these three parameters enter only in the same combination as that which represents the surface area of the black hole. One way to understand why is to recall the "area theorem": the event horizon area of a black hole cannot decrease; it increases in most transformations of the black hole. This increasing behavior is reminiscent of thermodynamic entropy of closed systems. Thus it is reasonable that the black hole entropy should be a monotonic function of area, and it turns out to be simplest such function.

If A stands for the surface area of a black hole (area of the event horizon), then the black hole entropy, in dimensionless form, is given by,

$$S_{BH} = \frac{A}{4L_p^2} = \frac{c^3 A}{4Gh}$$

where *LP* stands for the Planck length $G\hbar/c_3$ while G,$\hbar$ and c denote, respectively, Newton›s gravity constant, the Planck-Dirac constant ($h/(2\pi)$) and the speed of light. Of course, if the entropy in the usual (chemist's) form is required, the above should be multiplied by Boltzmann's constant k.

For the spherically symmetric and stationary, or Schwarzschild, black hole, the only parameter is the black hole's mass$M$, the horizon›s radius is $rh{=}2GM/c_2$, and its area is naturally given by $4\pi r_2h$, or

$$A = 16\pi(GM / c^2)^2$$

Note that a one-solar mass Schwarzschild black hole has a horizon area of the same order as the municipal area of Atlanta or Chicago. Its entropy is about 4×10^{77}, which are about twenty orders of magnitude larger than the thermodynamic entropy of the sun. This observation underscores the fact that one should not think of black hole entropy as the entropy that fell into the black hole when it was formed.

For the most general type of stationary black hole, the Kerr-Newman black hole (rotating black hole), the hole's parameters are mass M , electric charge Q and angular momentum J , and the horizon is no longer spherical. Nevertheless, in the popular Boyer-Lindquist coordinates $\{t,r,\theta,\varphi\}$ it lies at the fixed radial coordinate,

$$r = rh \equiv GM / c^2 + \sqrt{(GM / c^2)^2 - (G^{1/2}Q / c^2)^2 - (J / Mc)^2}$$

Consequently the horizon area is given by,

$$A = \int_0^{\pi} d\theta \int_0^{2\pi} d\varphi \sqrt{g_{\theta\theta} g_{\varphi\varphi}} = 4\pi(r_h^2 + (J / Mc)^2)$$

The First Law of Black Hole Thermodynamics

When near to equilibrium a thermodynamic system at temperature T changes its state, the consequent increments of its energy E and entropy S are related by the first law of thermodynamics:

$$TdS = dE - dW$$

Here dW is the work done on the system by exterior agents. When the system is one rotating with angular frequency Ω and charged up to electric potential Φ, the changes in its angular momentum J and charge Q contribute the work:

$$dW = \Omega dJ + \Phi dQ$$

to the above formula. A stationary black hole admits a similar relation. The differential dA from equation ($A = \int_0^{\pi} d\theta \int_0^{2\pi} d\varphi \sqrt{g_{\theta\theta} g_{\varphi\varphi}} = 4\pi(r_h^2 + (J / Mc)^2)$), when multiplied by a suitable factor, takes the form,

$$\Theta dA = d(Mc^2) - \Omega BH^{dJ} - \Phi_{BH} dQ$$

Where,

$$\Theta \equiv c^4 (2GA)^{-1}(r_h - GM / c^2)$$

$$\Omega_{BH} \equiv (J / Mc)(r_h^2 + (J / Mc)^2)^{-1}$$

$$\Phi_{BH} \equiv Q r_h (r_h^2 + (J / Mc)^2)^{-1}$$

Just now this is just a relation between increments in mechanical and geometrical properties. It turns out that Ω_{BH} is precisely the angular rotation frequency of the black hole in the sense that any test body dropped into it, as it approaches the horizon no matter where, ends up circumnavigating it at just this frequency. And Φ_{BH} turns out to be black hole›s electric potential in the sense that it equals the line integral of the hole's electric field from infinity to any location on the horizon.

Because Mc^2 is the hole's energy, equation ($\Theta dA = d(Mc^2) - \Omega BH^{dJ} - \Phi_{BH} dQ$) obviously looks like the first law for an ordinary thermodynamic system. It will be the first law if black hole entropy is required to be a function of A and of nothing else, so that $dS_{BH} \propto dA$. With the choice in equation (above) the black hole temperature T_{BH} must be,

$$T_{BH} = 4L_P^2\Theta = \frac{hc}{2\pi}\sqrt{\frac{(GM/c^2)^2 - (G^{1/2}Q^2/c^2)^2 - (J/Mc)^2}{r_h^2 + (J/Mc)^2}}$$

The reality of black hole temperature was brought home when Hawking showed that a non-eternal black hole must spontaneously emit thermal radiation with precisely this temperature the original calculation was for $J=0, Q=0$, but it is now clear that equation $T_{BH} = 4L_P^2\Theta = \frac{hc}{2\pi}\sqrt{\frac{(GM/c^2)^2 - (G^{1/2}Q^2/c^2)^2 - (J/Mc)^2}{r_h^2 + (J/Mc)^2}}$ is valid for all J and Q). This discovery provided the calibration of the numerical factor in equation (above).

The Generalized Second Law of Thermodynamics

In ordinary thermodynamics the second law requires that the entropy of a closed system shall never decrease, and shall typically increase as a consequence of generic transformations. While this law may hold good for a system including a black hole, it is not informative in its original form. For example, if an ordinary system falls into a black hole, the ordinary entropy becomes invisible to an exterior observer, so from her viewpoint, saying that ordinary entropy increases does not provide any insight: the ordinary second law is transcended.

Including the black hole entropy in the entropy ledger gives a more useful law, the generalized second law of thermodynamics (GSL): the sum of ordinary entropy So outside black holes and the total black hole entropy never decreases and typically increases as a consequence of generic transformations of the black hole. In equations the GSL extends the reach of the area theorem in both classical and quantum physics:

$$\Delta S_o + \Delta S_{BH} \geq 0$$

When matter entropy flows into a black hole, the GSL demands that the increase in black hole entropy shall more than compensate for the disappearance of ordinary entropy from sight. This has been verified by examples.

During the process of Hawking radiation the black hole›s area decreases (basically because of the decrement of black hole mass), in violation of the area theorem. This is known to reflect a failure of the energy condition (assumed by the theorem) as a result of the very quantum fluctuations which engender the radiation. The GSL predicts that the emergent Hawking radiation entropy shall more than compensate for the drop in black hole entropy. This has been verified amply, and stands as testament to the predictive power of the GSL which was formulated two years before the Hawking radiation was put in evidence. Varied theoretical arguments have been given in support of the GSL.

Status of the Third Law of Black Hole Thermodynamics

In ordinary thermodynamics the third law may be stated in two ways:

- Nernst-Simon statement: The entropy of a system at absolute zero temperature either vanishes or becomes independent of the intensive thermodynamic parameters, e.g. pressure, magnetic field, electric potential, etc.
- Unattainability statement: To bring a system to absolute zero temperature involves an infinite number of processes or steps.

From formula above it is clear that the black hole temperature vanishes when,

$$(GM / c^2)^2 - (G^{1/2}Q^2 / c^2)^2 - (J / Mc)^2 = 0$$

Kerr-Newman black holes satisfying this condition are called *extreme*. From equations above it is clear that for T_{BH}=0 the black hole entropy is not only nonvanishing, but depends on (J/Mc). Now this last quantity is an analog of a thermodynamic intensive parameter. For example, it is directly related to the hole's angular velocity in equation above, and angular velocity of a thermodynamic system is an intensive parameter. Thus the Nernst-Simon statement of the third law fails for black holes.

There is some evidence that the unattainability statement of the third law is satisfied by black holes. For example, in an astrophysical setting the process of spinning up a Q=0 black hole gets "hung up" at $(J/Mc)\approx 0.998GM/c^2$, before the externality condition can be satisfied.

What is Behind the Black Hole Entropy?

In ordinary statistical mechanics, the entropy S is a measure of the multiplicity of microstates that hide behind one particular macro state. A special case of this is Boltzmann›s famous formula $S=\ln W$ where W stands for the number of equally probable microstates of a particular macro state. Since black hole entropy plays a role quite analogous to that of ordinary entropy, e.g. it participates in the second law; many have wondered what the microstates that are counted by black hole entropy are. A partial list of interpretations of black hole entropy is the following:

- Black hole entropy counts the number of internal states of matter and gravity: As mentioned earlier, the perception that a particular black hole (specific M,J and Q) can be formed in many ways originally suggested the notion of black hole entropy; in this approach the internal states of matter and/or gravity are the sought for microstates. Examples of this viewpoint are provided by Frolov and Novikov and by Mukhanov.
- Black hole entropy is the entropy of entanglement between degrees of freedom inside and outside the horizon: For a black hole formed by collapse, the quantum field degrees of freedom external to the horizon should be entangled with those inside it. To an external observer the last are not accessible, so the meaningful state would be the one from which internal degrees of freedom have been traced out (removed). Even if the global state was pure and thus entropy free, this so reduced state will be a mixed one

and have entropy associated with it. Calculation shows that this entanglement entropy is proportional to the horizon area, just as required to explain black hole entropy, but the coefficient is ultraviolet divergent and thus requires renormalization by physical arguments.

- Black hole entropy counts the number of horizon gravitational states: It has been suggested that the sought for states are the states of the gravitational degrees of freedom residing on the black hole's horizon. An example of this approach is a calculation by Carlip based on the group of symmetries at the horizon. It reproduces formula above.

- Black hole entropy is a conserved quantity connected with coordinate invariance of the gravitational action: This abstract approach has been championed by Wald as a road to black hole entropy in more general theories of gravity than general relativity. Technically, black hole entropy is the Noether charge of the diffeomorphism symmetry. This notion reproduces formula above when the gravitational action is of first order in the curvature, but gives a modified formula for higher order gravity theories.

- Black hole entropy is thermal entropy of the gas of quanta constituting the thermal atmosphere of the black hole: The atmosphere concept comes from Thorne and Zurek and't Hooft; the last introduced the concept of a "brick wall" to keep the said atmosphere from contacting the horizon, and thus making the entropy infinite. This approach recovers the proportionality of entropy to horizon area, but the coefficient has to be chosen by hand.

- Black hole entropy counts the number of states or excitations of a fundamental string: Strings in string theory have a variety of excitations, so there is a multitude of string states. Therefore, a string has entropy, which turns out to be proportional to its *mass*. This is quite in contrast with black hole entropy. However, an argument by Bowick, Smolin and Wijewardhana suggests that by *adiabatically* (i.e. sufficiently slowly) reducing the string coupling constant g, it is possible to shrink a black hole's size as well as to reduce its mass (while keeping its entropy constant) until eventually it gets to be the size of the string length scale *ls* when the black hole should not be distinguishable from a string. At the corresponding value of g, string and black hole entropy are quite similar. This has been taken to mean that there is a one-to-one correspondence between black hole and string states, where both entities have the same entropy. This picture has been corroborated in the context of five-dimensional extreme black holes. Hence black hole entropy can be understood in terms of string entropy.

- Black hole entropy is equivalent to the thermal entropy of the radiation residing on the boundary of the space time containing the black hole: The AdS/CFT correspondence is a mapping between gravitational degrees of freedom of a certain space time and the matter (or field) degrees of freedom residing on its boundary. In particular, certain string theories in five dimensional Anti-deSitter (AdS) space time are so mapped to conformal field theories on the corresponding space time's four-dimensional boundary which bears some resemblance to Minkowski space time. Witten has shown that the entropy of a black hole residing in the bulk Anti-deSitter space time equals that of thermal radiation of the fields residing on its boundary.

Nuclear Astrophysics

Nuclear astrophysics is that branch of astrophysics which helps understanding the Universe, or at least some of its many faces, through the knowledge of the microcosm of the atomic nucleus. It attempts to find as many nuclear physics imprints as possible in the macrocosm, and to decipher what those messages are telling us about the varied constituent objects in the Universe at present and in the past. In the last decades much advance has been made in nuclear astrophysics thanks to the sometimes spectacular progress made in the modelling of the structure and evolution of the stars, in the quality and diversity of the astronomical observations, as well as in the experimental and theoretical understanding of the atomic nucleus and of its spontaneous or induced transformations. Developments in other sub-fields of physics and chemistry have also contributed to that advance. Notwithstanding the accomplishment, many longstanding problems remain to be solved, and the theoretical understanding of a large variety of observational facts needs to be put on safer grounds. In addition, new questions are continuously emerging, and new facts endangering old ideas.

Nuclear Astrophysics is a field at the intersection of nuclear physics and astrophysics that seeks to understand how nuclear processes shape the cosmos. In essence we look for the connection between properties of atomic nuclei and the properties of planets, stars, and galaxies. Open questions include "How did the universe create the elements?", "How can extremely dense and hot astrophysical environments be used to learn about fundamental properties of matter?", and "How is the energy created that powers stars and stellar explosions?" One fascinating aspect of this field is it's interdisciplinary and diversity. Work in nuclear astrophysics includes astronomical observations using telescopes, gravitational wave detectors, and neutrino detectors; accelerator laboratory experiments using beams of stable nuclei, radioactive nuclei, neutrons, and gamma-rays; laboratory analysis of interstellar grains; large scale computer simulations of stellar explosions and nuclei; and theoretical work in nuclear physics and astrophysics.

Nuclear astrophysics remains as a complex puzzle to science. The current consensus on the origins of elements and isotopes are that only hydrogen and helium (and traces of lithium, beryllium, boron) can be formed in a homogeneous Big Bang, while all other elements and their isotopes are formed in cosmic objects that formed later, such as in stars and their explosions.

The Sun's primary energy source is hydrogen fusion to helium at about 15 million degrees. The proton–proton chain reactions dominate, they occur at much lower energies although much more slowly than catalytic hydrogen fusion through CNO cycle reactions. Nuclear astrophysics gives a picture of the Sun's energy source producing a lifetime consistent with the age of the Solar System derived from meteoritic abundances of lead and uranium isotopes—an age of about 4.5 billion years. The core hydrogen burning of stars, as it now occurs in the Sun, defines the main sequence of stars, illustrated in the Hertzsprung-Russell diagram that classifies stages of stellar evolution. The Sun's lifetime of H burning via pp-chains is about 9 billion years. This primarily is determined by extremely slow production of deuterium,

$$ {}_1^1H + {}_1^1H \rightarrow {}_1^2D + e^+ + \nu_e + 0.42\,MeV $$

which is governed by the weak interaction. Work that led to discovery of neutrino oscillation (implying a non-zero mass for the neutrino absent in the Standard Model of particle physics) was motivated by a solar neutrino flux about three times lower than expected from theories—a long-standing concern in the nuclear astrophysics community colloquially known as the Solar neutrino problem.

The concepts of nuclear astrophysics are supported by observation of the element technetium (the lightest chemical element without stable isotopes) in stars, by galactic gamma-ray line emitters, by radioactive-decay gamma-ray lines from the 56Ni decay chain observed from two supernovae (SN1987A and SN2014J) coincident with optical supernova light, and by observation of neutrinos from the Sun and from supernova 1987a. These observations have far-reaching implications. 26Al has a lifetime of a million years, which is very short on a galactic timescale, proving that nucleosynthesis is an ongoing process within our Milky Way Galaxy in the current epoch.

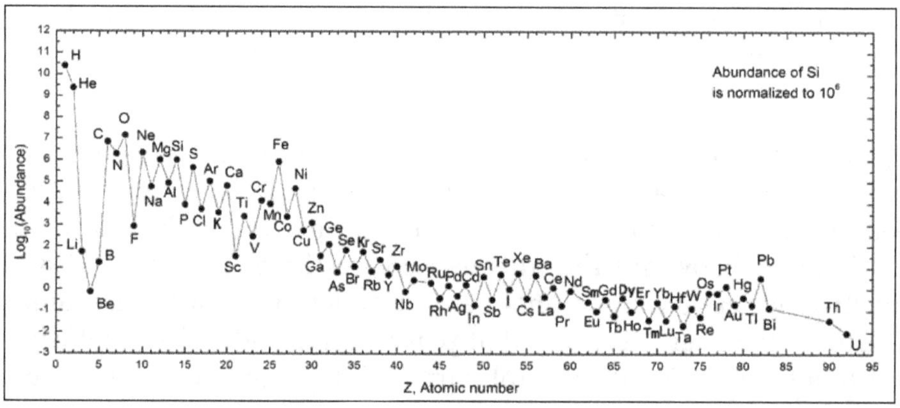

Abundances of the chemical elements in the Solar System. Hydrogen and helium are most common. The next three elements (Li, Be, B) are rare, intermediate-mass elements such as C, O, Si, Ca more abundant. Beyond Fe, there is a remarkable drop beyond Fe, heavier elements being 3-5 orders of magnitude less abundant. The two general trends in the remaining stellar-produced elements are: (1) an alternation of abundance of elements according to whether they have even or odd atomic numbers, and (2) a general decrease in abundance, as elements become heavier. Within this trend is a peak at abundances of iron and nickel, which is especially visible on a logarithmic graph spanning fewer powers of ten, say between logA=2 (A=100) and logA=6 (A=1,000,000).

Current descriptions of the cosmic evolution of elemental abundances are broadly consistent with those observed in the Solar System and galaxy, whose distribution spans twelve orders of magnitude (one trillion). The roles of specific cosmic objects in producing these elemental abundances are clear for some elements, and heavily debated for others. For example, iron is believed to originate mostly from thermonuclear supernova explosions (also called supernovae of type Ia), and carbon and oxygen is believed to originate mostly from massive stars and their explosions. Li, Be, and B are believed to originate from spallation reactions of cosmic-ray nuclei such as carbon and heavier nuclei, breaking these apart. Unclear is, in which sources nuclei much heavier than iron are produced; for the slow and rapid neutron capture reactions, different sites are discussed, such as envelopes of stars of either lower or higher masses, or supernova explosions versus collisions of compact stars. The transport of nuclear reaction products from their sources through the interstellar and intergalactic medium also is unclear, and there is, e.g., a missing metals problem of more production of heavy elements predicted than is observed in stars. Also, many nuclei that are involved in cosmic nuclear reactions are unstable and only predicted to exist temporarily in cosmic sites; we cannot easily measure the properties of such nuclei, and uncertainties on their binding energies are substantial. Similarly, stellar structure and its dynamics is not satisfactorily described in models and hard to observe except through aster seismology; also, supernova explosion models lack a consistent description based on physical processes, and include heuristic elements.

Computational Astrophysics

Computational astrophysics is the use of numerical methods to solve research problems in astrophysics on a computer. Numerical methods are used whenever the mathematical model describing an astrophysical system is too complex to solve analytically (with pencil and paper). Today, it is difficult to find examples of research problems that do not use computation.

Computational Versus Analytic Methods

Solutions generated by numerical methods are generally only approximations to the exact solution of the underlying equations. However, much more complex systems of equations can be solved numerically than can be solved analytically. Thus, approximate solutions to the exact equations found by numerical methods often provide far more insight than exact solutions to approximate equations that can be solved analytically. For example, time-dependent numerical solutions of fluid flow in three dimensions can exhibit behavior which is not expected from one-dimensional analytic solutions to the steady-state (time independent) equations.

The increase in computing power in the last few decades has meant that an increasingly larger share of problems in astrophysics can be solved on a desktop computer. However, the most computationally intensive problems in astrophysics are still limited by the memory and floating-point speed of the largest high-performance computer (HPC) systems available (e.g. computers on the top500 list). The use of HPC is necessary to maximize the spatial, temporal, or frequency resolution of solutions, to include more physics, or when large parameter surveys are required to understand the statistics.

Examples

Traditionally, the most important applications of computation in astrophysics have been in the areas of:

- Stellar structure and evolution: The internal structure and evolution of stars with different masses and chemical composition was largely mapped out in the 1960s using numerical methods to solve the equations of stellar structure. Today, the frontiers of research include calculating multidimensional stellar models of rapidly rotating stars, modeling the effects of hydro dynamical processes such as convection from first principles, and understanding how stars generate magnetic fields through dynamo processes.

- Radiation transfer and stellar atmospheres: Without oversimplifying assumptions, computational methods are required to calculate the propagation of light through the outer layers of a star, including its interaction with matter through absorption, emission, and scattering of photons. The calculation of cross sections for the interaction of light with matter for astrophysical relevant ions is itself a challenging computational problem. The construction of such stellar atmosphere models share many challenges with radiation transfer problems in other systems such as planets, accretion disks, and the interstellar medium. Modern calculations improve the frequency resolution, include a better treatment of opacities, and can treat non-hydrostatic atmospheres.

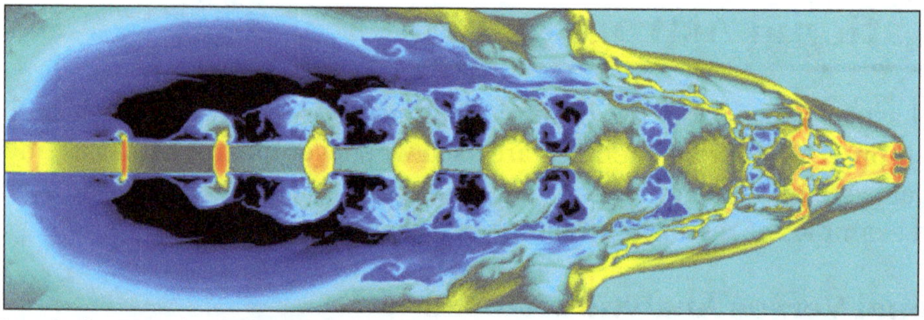

Density in a 2D simulation of the propagation of a pulsed, supersonic, protostellar jet.

- Astrophysical fluid dynamics: The dynamics of most of the visible matter in the universe can be treated as a compressible fluid. Time-dependent and multidimensional solutions to the fluid equations, including the effects of gravitational, magnetic, and radiation fields, require numerical methods. A vast range of problems are addressed in this way, from convection and dynamo action in stellar and planetary interiors, to the formation of galaxies and the large scale structure of the universe.

- Planetary, stellar, and galactic dynamics: It is well known that Newton's Laws of Motion for some number of particles N interacting through their mutual gravitational attraction do not in general have an analytic solution for $N > 2$. Thus, to compute the orbits of planets in the solar system, or of stars in the Galaxy, numerical methods are required. The most challenging problems today include accurate integration of the orbits of the planets over the age of the solar system, studying the dynamics of globular clusters including the effect of stellar evolution and the formation of binaries, studying galaxy mergers and interaction, and computing structure formation in the universe through the gravitational clustering of collisionless dark matter.

Numerical Methods

The diverse set of mathematical models encountered in astrophysics means that a very wide range of numerical methods are necessary. These range from basic methods for linear algebra, nonlinear root finding, and ordinary differential equations (ODEs), to more complex methods for coupled partial differential equations (PDEs) in multi-dimensions, topics which span the entire content of reference books such as Numerical Recipes.

However, there are several numerical methods used in astrophysics that deserve special mention, either because they have such wide use in astrophysics, or because astrophysicists have made significant contributions to their development. These methods are considered in the following subsections.

Stellar Structure Codes

The equations of stellar structure are a set of ODEs which define a classic two-point boundary value problem. Analytic solutions to an approximate system of equations exist in special cases (polytropes) but these are of limited application to real stars. Numerical solutions to the full system of equations were first computed using shooting methods, in which boundary conditions are guessed at the center and surface of the star, the equations are integrated outwards and inwards and matching conditions are used at some interior point to select the full solution. Shooting methods are laborious and cumbersome; today modern stellar structure codes use relaxation schemes which find the solution to the finite-difference form of the stellar structure equations by finding the roots of coupled, non-linear equations at each mesh point. A good example of a public code that uses a relaxation scheme is the EZ-code, based on Eggleton's variable mesh method. The evolutions of stars are then computed by computing stellar models at discrete time intervals, with the chemical composition of the star modified by nuclear reactions in the interior.

Radiative Transfer Codes

Calculating the emergent intensity from an astrophysical system requires solving a multidimensional integral-differential equation, along with level population equations to account for the interaction of the radiation with matter. In general the solution is a function of two angles, frequency, and time. Even in static, plane-parallel atmospheres, the problem is two-dimensional (one angle and frequency). However, the most challenging aspect of the problem is that scattering couples the solutions at different angles and frequencies. As in the stellar structure problem, relaxation schemes are used to solve the finite difference form of the transfer equations, although specialized iteration techniques are necessary to accelerate convergence. Monte-Carlo methods, which adopt statistical techniques to approximate the solution by following the propagation of many photon packets, are becoming increasingly important. The problem of line-transfer in a moving atmosphere (stellar wind) is especially challenging, due to non-local coupling introduced by Doppler shifts in the spectrum.

N-body Codes

There are two tasks in an N-body code: integrating the equations of motion (pushing particles), and computing the gravitational acceleration of each particle. The former requires methods for integrating ODEs. Modern codes are based on a combination of high-order difference approximations (e.g. Hermite integrators), and symplectic methods (which have the important property of generating solutions that obey Liouville's Theorem, i.e. that preserve the volume of the solution in phase space). Symplectic methods are especially important for long term time integration, because they control the accumulation of truncation error in the solution.

Calculation of the gravitational acceleration is challenging because the computational cost scales as $N (N^{-1})$, where N is the number of particles. For small N, direct summation can be used. For moderate N (currently $N \sim 10^{5-6}$), special purpose hardware (e.g. GRAPE boards) can be used to accelerate the evaluation of $1/r^2$ necessary to compute the acceleration by direct summation. Finally, for large N (currently $N \geq 10^{9-10}$), tree-methods are used to approximate the force from distant particles.

Codes for Astrophysical Fluid Dynamics

Solving the equations of compressible gas dynamics is a classic problem in numerical analysis which has application to many fields besides astrophysics. Thus, a large number of methods have been developed, with many important contributions being made by astrophysicists. A complete review of all the methods is beyond the scope of this discussion. For solving the equations of compressible fluid dynamics, the most popular methods include:

- Finite-difference techniques (which require hyper-viscosity to smooth discontinuities),
- Finite-volume methods (which often use a Riemann solver to compute upwind fluxes),
- Operator split methods (which combine elements of both finite-differencing and finite-volume methods for different terms in the equations),
- Central methods (which often use simple expressions for the fluxes, combined with high-order spatial interpolation),
- Particle methods such as smooth particle hydrodynamics (SPH, which integrates the motion of discrete particles to follow the flow).

A good technical review of many of these methods is given by LeVeque. SPH is an example of a method developed largely to solve astrophysics problems, although many of the developments in other methods (e.g. the extension of finite-difference and finite-volume methods to magnetohydrodynamics (MHD) to include the effects of magnetic fields on the dynamics of the fluid) have also been motivated by astrophysics.

3

Theories of Astrophysics

Astrophysics has numerous theories associated with it. These include the Big Bang which proposes the creation of the universe and its ever expanding nature or the Partial Impact Theory which describes the creation a third star through the partial collision of two stars fall within its aspect. This chapter aims to shed light on the various astrophysics theories to provide a better understanding.

Big Bang Theory

Big Bang theories are fairly simple. In short, the Big Bang hypothesis states that all of the current and past matter in the Universe came into existence at the same time, roughly 13.8 billion years ago. At this time, all matter was compacted into a very small ball with infinite density and intense heat called a Singularity. Suddenly, the Singularity began expanding and the universe as we know it began.

While this is not the only modern theory of how the Universe came into being for example, there is the Steady State Theory or the Oscillating Universe Theory it is the most widely accepted and popular. Not only does the model explain the origin of all known matter, the laws of physics, and the large scale structure of the Universe, it also accounts for the expansion of the Universe and a broad range of other phenomena.

Working backwards from the current state of the Universe, scientists have theorized that it must have originated at a single point of infinite density and finite time that began to expand. After the initial expansion, the theory maintains that Universe cooled sufficiently to allow the formation of subatomic particles, and later simple atoms. Giant clouds of these primordial elements later coalesced through gravity to form stars and galaxies.

This all began roughly 13.8 billion years ago, and is thus considered to be the age of the universe. Through the testing of theoretical principles, experiments involving particle accelerators and high-energy states, and astronomical studies that have observed the deep universe, scientists have constructed a timeline of an event that began with the Big Bang and has led to the current state of cosmic evolution.

However, the earliest times of the Universe lasting from approximately 10^{-43} to 10^{-11} seconds after the Big Bang are the subject of extensive speculation. Given that the laws of physics as we know them could not have existed at this time, it is difficult to fathom how the Universe could have been governed. What's more, experiments that can create the kinds of energies involved have not yet been conducted. Still, many theories prevail as to what took place in this initial instant in time, many of which are compatible.

Singularity Epoch

Also known as the Planck Epoch (or Planck Era), this was the earliest known period of the Universe. At this time, all matter was condensed on a single point of infinite density and extreme heat. During this period, it is believed that the quantum effects of gravity dominated physical interactions and that no other physical forces were of equal strength to gravitation.

This Planck period of time extends from point 0 to approximately 10^{-43}seconds, and is so named because it can only be measured in Planck time. Due to the extreme heat and density of matter, the state of the universe was highly unstable. It thus began to expand and cool, leading to the manifestation of the fundamental forces of physics.

From approximately 10^{-43} second and 10^{-36}, the universe began to cross transition temperatures. It is here that the fundamental forces that govern the Universe are believed to have begun separating from each other. The first step in this was the force of gravitation separating from gauge forces, which account for strong and weak nuclear forces and electromagnetism.

Then, from 10^{-36} to 10^{-32} seconds after the Big Bang, the temperature of the universe was low enough (10^{28} K) that the forces of electromagnetism (strong force) and weak nuclear forces (weak interaction) were able to separate as well, forming two distinct forces.

Inflation Epoch

With the creation of the first fundamental forces of the universe, the Inflation Epoch began, lasting from 10^{-32} seconds in Planck time to an unknown point. Most cosmological models suggest that the Universe at this point was filled homogeneously with a high-energy density, and that the incredibly high temperatures and pressure gave rise to rapid expansion and cooling.

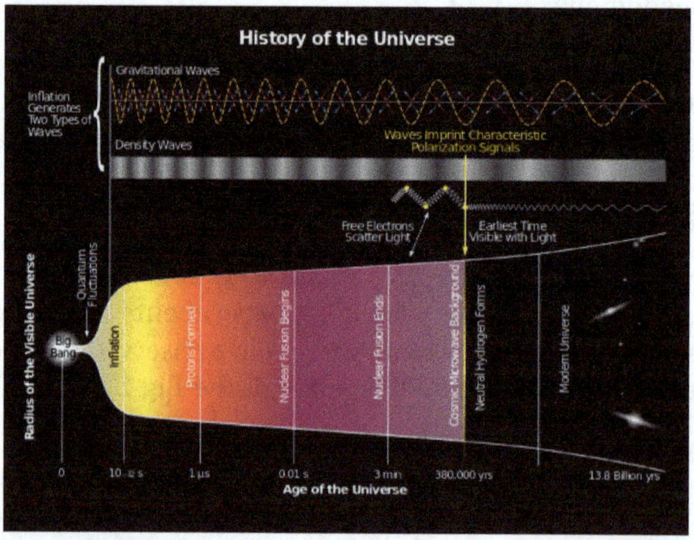

The history of the Universe, from the Big Bang to the current epoch.

This began at 10^{-37} seconds, where the phase transition that caused for the separation of forces also led to a period where the universe grew exponentially. It was also at this point in time that baryogenesis occurred, which refers to a hypothetical event where temperatures were so high that the random motions of particles occurred at relativistic speeds.

As a result of this, particle–antiparticle pairs of all kinds were being continuously created and destroyed in collisions, which is believed to have led to the predominance of matter over antimatter in the present universe. After inflation stopped, the universe consisted of quark–gluon plasma, as well as all other elementary particles. From this point onward, the Universe began to cool and matter coalesced and formed.

Cooling Epoch

As the universe continued to decrease in density and temperature, the energy of each particle began to decrease and phase transitions continued until the fundamental forces of physics and elementary particles changed into their present form. Since particle energies would have dropped to values that can be obtained by particle physics experiments, this period onward is subject to less speculation.

For example, scientists believe that about 10^{-11} seconds after the Big Bang, particle energies dropped considerably. At about 10^{-6} seconds, quarks and gluons combined to form baryons such as protons and neutrons, and a small excess of quarks over antiquarks led to a small excess of baryons over antibaryons.

Since temperatures were not high enough to create new proton-antiproton pairs (or neutron-anitneutron pairs), mass annihilation immediately followed, leaving just one in 10^{10} of the original protons and neutrons and none of their antiparticles. A similar process happened at about 1 second after the Big Bang for electrons and positrons. After these annihilations, the remaining protons, neutrons and electrons were no longer moving relativistically and the energy density of the universe was dominated by photons and to a lesser extent, neutrinos.

A few minutes into the expansion, the period known as Big Bang nucleosynthesis also began. Thanks to temperatures dropping to 1 billion kelvin and the energy densities dropping to about the equivalent of air, neutrons and protons began to combine to form the universe's first deuterium (a stable isotope of Hydrogen) and helium atoms. However, most of the Universe's protons remained uncombined as hydrogen nuclei.

After about 379,000 years, electrons combined with these nuclei to form atoms (again, mostly hydrogen), while the radiation decoupled from matter and continued to expand through space, largely unimpeded. This radiation is now known to be what constitutes the Cosmic Microwave Background (CMB), which today is the oldest light in the Universe.

As the CMB expanded, it gradually lost density and energy, and is currently estimated to have a temperature of 2.7260 ± 0.0013 K (-270.424 °C/ -454.763 °F) and an energy density of 0.25 eV/cm^3 (or 4.005×10^{-14} J/m^3; 400–500 photons/cm^3). The CMB can be seen in all directions at a distance of roughly 13.8 billion light years, but estimates of its actual distance place it at about 46 billion light years from the center of the Universe.

Structure Epoch

Over the course of the several billion years that followed, the slightly denser regions of the almost uniformly distributed matter of the Universe began to become gravitationally attracted to each other. They therefore grew even denser, forming gas clouds, stars, galaxies, and the other astronomical structures that we regularly observe today.

This is what is known as the Structure Epoch, since it was during this time that the modern Universe began to take shape. This consists of visible matter distributed in structures of various sizes, ranging from stars and planets to galaxies, galaxy clusters, and super clusters where matter is concentrated that are separated by enormous gulfs containing few galaxies.

The details of this process depend on the amount and type of matter in the universe, with cold dark matter, warm dark matter, hot dark matter, and baryonic matter being the four suggested types. However, the Lambda-Cold Dark Matter model (Lambda-CDM), in which the dark matter particles moved slowly compared to the speed of light is the considered to be the standard model of Big Bang cosmology, as it best fits the available data.

In this model, cold dark matter is estimated to make up about 23% of the matter/energy of the universe, while baryonic matter makes up about 4.6%. The Lambda refers to the Cosmological Constant, a theory originally proposed by Albert Einstein that attempted to show that the balance of mass-energy in the universe was static. In this case, it is associated with Dark Energy, which served to accelerate the expansion of the universe and keep its large-scale structure largely uniform.

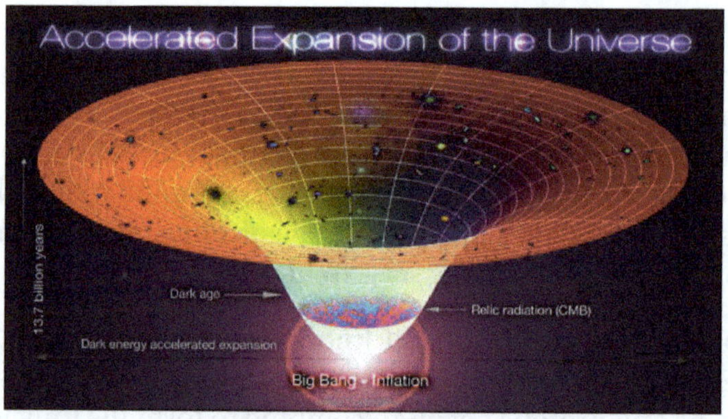

Diagram showing the Lambda-CBR universe, from the Big Bang to the the current era.

Long-term Predictions for the Future of the Universe

Hypothesizing that the Universe had a starting point naturally gives rise to questions about a possible end point. If the Universe began as a tiny point of infinite density that started to expand, does that mean it will continue to expand indefinitely? Or will it one day run out of expansive force, and begin retreating inward until all matter crunches back into a tiny ball?

Answering this question has been a major focus of cosmologists ever since the debate about which model of the Universe was the correct one began. With the acceptance of the Big Bang Theory, but prior to the observation of Dark Energy in the 1990s, cosmologists had come to agree on two scenarios as being the most likely outcomes for our Universe.

In the first, commonly known as the "Big Crunch" scenario, the universe will reach a maximum size and then begin to collapse in on it. This will only be possible if the mass density of the Universe is greater than the critical density. In other words, as long as the density of matter remains at or above a certain value ($1\text{-}3 \times 10^{-26}$ kg of matter per m^3), the Universe will eventually contract.

Alternatively, if the density in the universe were equal to or below the critical density, the expansion would slow down but never stop. In this scenario, known as the "Big Freeze", the Universe would go on until star formation eventually ceased with the consumption of all the interstellar gas in each galaxy. Meanwhile, all existing stars would burn out and become white dwarfs, neutron stars, and black holes.

Very gradually, collisions between these black holes would result in mass accumulating into larger and larger black holes. The average temperature of the universe would approach absolute zero, and black holes would evaporate after emitting the last of their Hawking radiation. Finally, the entropy of the universe would increase to the point where no organized form of energy could be extracted from it (a scenarios known as "heat death").

Modern observations, which include the existence of Dark Energy and its influence on cosmic expansion, have led to the conclusion that more and more of the currently visible universe will pass beyond our event horizon (i.e. the CMB, the edge of what we can see) and become invisible to us. The eventual result of this is not currently known, but "heat death" is considered a likely end point in this scenario too.

Other explanations of dark energy, called phantom energy theories, suggest that ultimately galaxy clusters, stars, planets, atoms, nuclei, and matter itself will be torn apart by the ever-increasing expansion. This scenario is known as the "Big Rip", in which the expansion of the Universe itself will eventually be its undoing.

Vanishing Dimensions Theory

Despite the fascinating success of the modern physics embodied in the Standard Model of particle physics and the Standard Model of Cosmology, serious problems have accumulated that need prompt attention. It is very difficult to imagine further model building in both particle physics and cosmology without addressing fundamental problems like the Standard Model hierarchy problem, the problem of initial conditions in cosmology, the cosmological constant problem, the dark matter problem, the black hole information loss problem etc. It is becoming increasingly clear that straightforward extensions of existing theories will not suffice and some radically new ideas are needed. Here, we propose a new approach. Instead of changing the mathematical details of the well-established physical theories, we can change the background on which these theories are formulated. This shift will create enough room for looking at the many fundamental problems from a completely new angle.

Most of the fundamental problems that we are currently facing stem from the ultraviolet (short distance) and infrared (large distance) divergences. We believe that (with some exceptions) we understand our universe on scales approximately between 10^{-17} cm and a G_{pc}. The first scale corresponds to the energy scale of TeV^{-1} which is the energy probed in the highest energy accelerators available so far. The second scale corresponds to the distance characteristic for super-clusters of galaxies, i.e. the scale at which cosmology kicks in. At scales shorter than 10^{-17} cm and larger than about a G_{pc} we are running into problems. At the intermediate scales between 10^{-17} cm and a G_{pc} we know pretty well that our space is three dimensional. We propose that the space at scales shorter than 10^{-17} cm is lower dimensional, while at scales larger than a G_{pc} the space is higher dimensional. The most natural setup in this sense would be the one in which the effective number of dimensions increases with the length scale. At the fundamental level, the space-time is probably quantized, i.e. discrete. We leave aside the problem of time and consider only the evolution of space. At the shortest distances at which our space appears as continuum, the space is one-dimensional. At a certain critical length scale, the space becomes effectively two-dimensional. At the scale of about 10^{-17} cm, the space becomes effectively three-dimensional. Finally, at the scales of about a G_{pc}, the space becomes effectively four-dimensional. In principle, this hierarchy does not need to stop a priori at any finite number. However, it would be interesting for other reasons if this construct stops at 10 or 26 effective dimensions. In a dynamical picture where the universe starts from zero size and then grows, there is no background with the fixed number of dimensions in which the universe expands. The expanding structure actually builds the effective dimensions during its evolution.

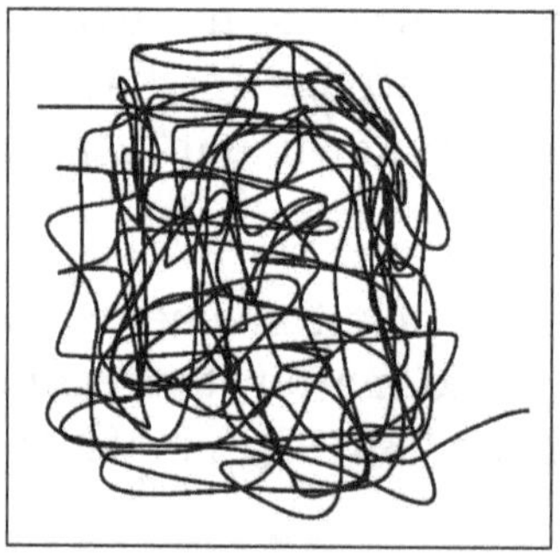

An example of a structure which is one-dimensional on short scales while it appears effectively two-dimensional at large scales.

An evolving background with such features can be effectively described with a formalism of an induced metric on a sub-manifold embedded in a higher dimensional manifold. The induced metric can be written as $\gamma_{ab} = g_{\mu\upsilon}\partial_a X^{\mu}\partial_b X^{\upsilon}$ where $g_{\mu\nu}$ is the background metric on a higher dimensional manifold, Latin indices cover the induced metric, while Greek indices cover the background metric. Usually, $g_{\mu\nu}$ is considered to be the fundamental metric of the space time, while γ_{ab} is the induced one on the sub-manifold. In our approach, the roles are inverted. γ_{ab} is the fundamental metric since it builds the effective foreground (rather than background) metric $g_{\mu\nu}$. The model formulated in (with quite a different motivation) can be nicely used for this purpose. In that model, a string with the manifest tendency to grow was considered. The string was chopped into N segments, each carrying longitudinal momentum P^+/N, where P^+ is the total longitudinal momentum. The Hamiltonian for such a string is given by,

$$H = \sum_{i}^{N} \frac{2}{2P^{+}(i)} \left\{ \vec{P}_{\perp}^{2}(i) + \left[\vec{X}(i-1) - \vec{X}(i) \right]^{2} \right\}$$

where P~⊥(i) and X~ (i) are the transverse momentum and position of the i-th segment. The non-vanishing longitudinal momentum causes the string the grow in length, while the non-vanishing transverse momentum is the source of the string's extrinsic curvature. Thus, such a string builds a structure described in Fig. It can be shown that the total length of the string grows as N, while the radius of the effective space covered by such a string grows as $\sqrt{\log N}$. In the limit of $N \to \infty$ the string becomes space-filling.

Curing ultraviolet divergences. There exist a strong motivation for changing the dimensionality of the space-time at small distances. One of the most acute problems connected with ultraviolet divergences is the Standard Model hierarchy problem. The Higgs Lagrangian together with Yukawa couplings in the Standard Model is,

$$L_H = D_\nu \Phi^\dagger D^\nu \Phi - \mu^2 \Phi^\dagger \Phi + \frac{\lambda}{2}\left(\Phi^\dagger \Phi\right)^2 - \sum_f g_f \Phi \bar{\psi}_f \psi_f$$

where Φ is the Higgs field, $g_f^2 = m_f^2 / v^2, \lambda = m_H^2 / \left(2v^2\right)$, m_f and m_H are the fermion and Higgs mass respectively, while v is the Higgs field vacuum expectation value. If we consider all 1 loop one-particle-irreducible diagrams, then we find that the Higgs self-energy comes from the three types of diagrams. In 3 + 1 dimensions, all of these terms are quadratic ally divergent with the cut-off scale Λ at which new physics appears. The contribution of fermions, gauge bosons and the Higgs itself in the loop are respectively.

$$i\frac{g_f^2}{2}\int^\Lambda \frac{d^4k}{(2\pi)^4} tr\left(\frac{i}{\not{k} - m_f}\frac{i}{\not{k} + \not{p} - m_f}\right) \sim \Lambda^2 \frac{g_f^2}{32\pi^2}$$

$$i\frac{g^2}{2}\int^\Lambda \frac{d^4k}{(2\pi)^4}\frac{1}{k^2 - m_g^2} \sim \Lambda^2 \frac{g_f^2}{64\pi^2}$$

$$i3\lambda\int^\Lambda \frac{d^4k}{(2\pi)^4}\frac{1}{k^2 - m_H^2} \sim \Lambda^2 \frac{g_f^2}{16\pi^2}$$

Here, g is the gauge coupling constant, $g^2 = 2m_g^2 / v^2$, while m_g is the mass of gauge boson. The quadratic divergence implies that a very special cancelation needs to happen between the bare Higgs mass and the corrections, unless the scale of new physics Λ is very close to the electroweak scale.

Solutions to the hierarchy problem proposed so far postulate new physics not very far from the electroweak scale. An alternative approach would be to keep the Standard Model physics, and change the dimensionality of the background on which the model is defined. For example, in 2 + 1 dimensional space-time all of the terms are only linearly divergent.

$$i\frac{g_f^2}{2}\int^\Lambda \frac{d^3k}{(2\pi)^3} tr\left(\frac{i}{\not{k} - m_f}\frac{i}{\not{k} + \not{p} - m_f}\right) \sim -\Lambda^2 \frac{g_f^2}{4\pi^2}$$

$$\frac{g^2}{2}\int^\Lambda \frac{d^3k}{(2\pi)^3}\frac{1}{k^2 - m_g^2} \sim \Lambda \frac{g_f^2}{8\pi^2}$$

$$i3\lambda\int^\Lambda \frac{d^3k}{(2\pi)^3}\frac{1}{k^2 - m_H^2} \sim \Lambda \frac{3\lambda}{2\pi^2}$$

Going further, in 1 + 1 dimensional space-time all of these terms are only logarithmically divergent,

$$i\frac{g_f^2}{2}\int^\Lambda \frac{d^2k}{(2\pi)^2} tr\left(\frac{i}{\not{k} - m_f}\frac{i}{\not{k} + \not{p} - m_f}\right) \sim -\log(\Lambda / m_f)\frac{g_f^2}{4\pi}$$

$$\frac{g^2}{2}\int^{\Lambda}\frac{d^2k}{(2\pi)^2}\frac{1}{k^2-m_g^2}\sim\log(\Lambda/m_f)\frac{g^2}{8\pi}$$

$$i3\lambda\int^{\Lambda}\frac{d^2k}{(2\pi)^2}\frac{1}{k^2-m_H^2}\sim\log\left(\Lambda/m_f\right)\frac{3\lambda}{2\pi^2}$$

Thus, keeping the Standard Model Lagrangian and lowering the dimensionality of the space-time greatly improves the fine tuning problem in the Standard Model. In fact the dimensional regularization procedure tells us that the ultraviolet divergences in field theory are poles in the dimension plane. Lowering the dimensionality of the space-time universally cures ultraviolet divergences in practically all of the field theories.

Possible implications of having less dimensions at higher energies are very important for the LHC physics. There are three immediate and spectacular consequences of this model at the LHC, which should be observable if the dimensional crossover scale is ~ 1 TeV, i.e., within the reach of the machine: (i) cross section of hard scattering processes changes compared to that in the SM as the Q^2 becomes comparable with the crossover scale; (ii) 2 → 4 and higher order scattering processes at high energies become planar, resulting, e.g., in four-jet events, where all jets are produced in one plane in their center-of-mass frame, thus strikingly different from standard QCD multijet events; (iii) under certain conditions, jets of sufficiently high energy may become elliptic in shape.

What about gravity? The most elusive concept in modern physics – quantum gravity - is much more within the reach in lower dimensions. Gravity in 3+ 1 dimensions is complicated, highly nonlinear, perturbatively non-renormalizable theory. All of the attempts to quantize gravity in 3+ 1 dimensions failed so far. However, we realize that if the fundamental short scale physics is lower dimensional, there is no need to quantize 3 + 1 dimensional gravity. Instead we should quantize 2 + 1 and 1 + 1 dimensional gravity. This is much easier task to accomplish. In any space-time, the curvature tensor $R_{\mu\nu\rho\sigma}$ may be decomposed into a Ricci scalar R, Ricci tensor $R_{\mu\nu}$ and conformally invariant Weyl tensor $C_{\sigma\ \mu\nu\rho}$. In 2 + 1 dimensions the Weyl tensor vanishes and $R_{\mu\nu\rho\sigma}$ can be expressed solely through $R_{\mu\nu}$ and R. Explicitly,

$$R_{\mu\nu\rho\sigma}=\varepsilon_{\mu\nu\alpha}\varepsilon_{\rho\sigma\beta}G^{\alpha\beta}$$

This in turn implies that any solution of the vacuum Einstein's equations is locally flat. Thus, 2 + 1 dimensional space-time has no local gravitational degrees of freedom, i.e. no gravitational waves in classical theory and no gravitons in quantum theory. The number of degrees of freedom in such a theory is finite, quantum field theory reduces to quantum mechanics and the problem of non-renormalizability disappears. Obviously, 2 + 1 dimensional gravity has much nicer structure than its 3 + 1 dimensional cousin. For the reason of simplicity, 1 + 1 dimensional gravity is even more attractive. Einstein's action in 1 + 1 dimensional space-time is a constant (Euler's characteristic of the manifold in question) and the theory is trivial (unless augmented by some additional fields). Models of gravity in 1 + 1 dimensions are completely solvable and a lot of work has been done on their quantum aspects.

If the space-time is 2 + 1 dimensional at distances shorter than 10^{-17} cm, then we expect some strong implications for the high energy scattering processes. The structure of well-established theories like Quantum Chromo Dynamics (QCD) becomes much simpler in 2 + 1 and 1 + 1 dimensional space-times. The form of the QCD Lagrangian in 2 + 1 dimensions is the same as in 3 + 1 dimensions,

$$L = -\frac{1}{4} F^a_{\mu\nu} F^{a\mu\nu} + i\bar{\psi}\gamma^\mu \left(\partial_\mu + igA^a_\mu T^a\right)\psi$$

$$F^a_{\mu\upsilon} = \partial_\mu A^a_\upsilon - \partial_\nu A^a_\upsilon + gf^{abc} A^b_\mu A^b_\mu$$

except that the γ-matrices can be chosen to be proportional to two-dimensional Pauli matrices, while spinors ψ are two-component spinors. It is interesting that 2 + 1 dimensional QCD is super-renormalizable, i.e. only a finite set of graphs need overall counter terms. This is a consequence of the fact that the coupling constant in this theory has positive dimension. In 2 + 1 dimensional space-time there is only one transverse dimension, so there is no arbitrarily high transverse angular momentum. This implies that there exists no Regge-like behavior due to exchange of states of high spin which is characteristic in 3 + 1 dimensional space-time. For the LHC, it is certainly very important to calculate hadron-hadron scattering amplitude. It is not difficult to verify that the result is quite different from the standard one. The total cross section falls off like 1/log s, where s is the center of mass energy squared. Characteristic Regge factor $s^{\alpha}, \alpha > 0$ is completely absent , in the strong contrast with 3 + 1 dimensions.

For completeness, we mention 1 + 1 dimensional QCD. QCD in 1 + 1 dimensions is trivially asymptotically free, being super-renormalizable. The model on the infinite line has no gluon degrees of freedom, but it is a self-interacting fermion theory. In QCD on a circle, boundary conditions force the retention of quantum mechanical (zero-mode) gauge degrees of freedom in a Hamiltonian formulation. The dynamics of the zero modes lead to an elimination of fermionic non-singlet states from the spectrum in the continuum limit , thus practically eliminating color. While the details of 1+ 1 and especially 2+ 1 dimensional QCD are very interesting for collider phenomenology, they cannot help (at least not directly) with the problem of quark confinement in a nucleon since the effective size of a nucleon (GeV^{-1}) is much larger than the critical distance at which the space appears 3 + 1 dimensional (TeV^{-1}).

Curing Infrared Divergences

From the dimensional regularization point of view, infrared divergences, like their cousins ultraviolet ones, are also poles in the dimension plane. Therefore, changing the effective dimensionality of the space-time will remove them trivially. For example, the electron-photon interaction cross section is divergent at low momenta (large wavelengths) of the virtual photon. The probability amplitude P for this process is:

$$P \approx \frac{a}{\pi} \int_0^{|q|} dk \frac{1}{k} I(\upsilon, \upsilon')$$

Here, k is the momentum of the virtual photon; |q| is its maximal value, while v and v ′ are initial and final velocity of the electron. $I(\upsilon,\upsilon') = dE / dk$ Is the intensity, which is independent of k when k is small? The probability is obviously divergent at low k. We can remove this divergence by introducing an infrared momentum cutoff at very low energies. This can be ad-hock justified as a limitation of a finite size detector. However, changing the effective dimensionality of the space-time at large scales also removes this divergence. In our context, if the space-time becomes effectively 4 + 1 dimensional the integral becomes,

$$P \propto \int_0^{|q|} dk$$

This integral is not divergent as k → 0 and the infrared divergence is cured. Thus, increasing the dimensionality of the space-time effectively cures infrared divergences in field theories.

Changing the dimensionality of the space-time at large distances may have some consequences for cosmology. This can happen if the length scale at which the fourth spatial dimension opens up is smaller than the present cosmological horizon. Available observational data indicate that our universe is going through a phase of accelerated expansion. To date it remains a mystery what is the driving force behind the acceleration. Data favor an equation of state of the cosmic fluid $p = -\rho$, corresponding to a constant energy density. The null hypothesis is that we are observing the action of the vacuum energy density (or cosmological constant). If it is indeed the cosmological constant we are seeing, it may represent the worst prediction ever made by a theory. Instead, the cosmological constant may just be a shadow that extra dimensions cast on our visible universe. To demonstrate this, we write down the metric of 4+ 1 dimensional space-time (with spatially isotropic 3 dimensional slices) as,

$$ds^2 = e^{\nu}dt^2 - e^{\omega}\left(dr^2 + r^2 d\Omega^2\right) - e^{\mu}d\psi^2$$

where ψ is the fourth spatial dimension, $d\Omega^2 \equiv d\theta^2 + \sin^2\theta d\varphi^2$, while parameters ν,μ and ω are arbitrary functions of t and ψ. We can write down 4 + 1 dimensional vacuum Einstein's equations,

$$G_{AB} \equiv R_{AB} - \frac{1}{2} g_{AB}R = 0$$

where indices A and B go over all of the five coordinates. One of the homogeneous and isotropic solutions of these vacuum Einstein's equations found in deserves special attention:

$$ds^2 = dt^2 - e^2\sqrt{\Lambda/3t}\ (dr^2 + r^2 d\Omega^2) - d\psi^2$$

where $\Lambda = 3/\psi^2$. This metric reduces on ψ = constant hypersurfaces to a 3 + 1 dimensional de Sitter metric with Λ = constant. An observer located on a ψ = constant slice of the space-time will measure the effective stress energy tensor,

$$8\pi G T_{\mu\nu} = G_{\mu\nu}$$

where the Greek indices go over 3 + 1 dimensional hypersurface only. Note that these equations are already contained in $G_{AB} = 0$. Thus, the"matter" described by $T_{\mu\nu}$ is a manifestation of pure geometry in the higher dimensional space-time and it was called"induced matter" in or"shadow matter" in. The equation of state of matter defined with this stress energy tensor is $p = -\rho$

with $\rho = \Lambda/(8\pi G)$. This solution would not be of much use in theories with compact or warped extra dimensions where the effective size of extra dimensions is small, or brane world models with non-zero tension where $T_{\mu\nu}$ would get extra contributions. However, in our framework, where the three-dimensional sheet we are located on is embedded into a large structure which is effectively 4 + 1 dimensional on cosmological distances, full advantage of this solution can be used. In this framework, we can address the question of the smallness of the observed cosmological constant in a completely different way. We see that observers located at different slices of five-dimensional space-time infer different values of the effective cosmological constant. The small value of the cosmological constant that we observe can be attributed to the position of our 3 + 1 dimensional slice in the full 4 + 1 dimensional bulk. Along these lines one might argue that small values of Λ (i.e. large values of ψ) are much more natural than the opposite. Indeed, the vacuum energy density of $\rho = (10^{-3}\,eV)^4$ corresponds to the numerical value of $\psi \sim 10^{60} M^{-1}_{Pl}$. This is comparable to the current horizon size, which is in this scenario comparable to the characteristic distance between 3 + 1 dimensional sheets comprising a 4 + 1 dimensional structure we live in.

What makes this proposal of evolving dimensions very attractive is that some evidence of the lower dimensional structure of our space-time at a TeV scale may already exist. Namely, alignment of the main energy fluxes in a target (transverse) plane has been observed in families of cosmic ray particles. The fraction of events with alignment is statistically significant for families with energies higher than TeV and large number of hadrons. This can be interpreted as evidence for coplanar scattering of secondary hadrons produced in the early stages of the atmospheric cascade development.

An interesting side-effect of such a dimensional reduction scheme is the distinct nature of gravity in lower dimensions. It is well-known that, in a (2+ 1)-dimensional Einstein's universe, there are no local gravitational degrees of freedom, and hence there are no gravitational waves (or gravitons). If the universe was indeed (2 + 1)- dimensional at some earlier epoch, it is reasonable to deduce that no primordial gravitational waves (PGWs) of this era exist today. There is thus a maximum frequency for PGWs, implicitly related to the dimensional transition scale Λ_2, beyond which no waves can exist. This indicates that gravitational wave astronomy can be used as a tool for probing the novel "vanishing dimensions" framework.

Partial Impact Theory

Partial impact theory is an astronomical theory describing the partial collision of two stars and the temporary creation of a bright third star as a consequence. It is not part of contemporary astrophysics. A slight "grazing" collision between stars would be much more common than a head on impact between stars. So he believed this phenomenon needed to be explained to account for the appearance of bright new stars that would appear in the night sky and disappear within a year or even days.

The theory explains that when the two stellar bodies graze each other, the grazed parts will shear off from the main body of each star. Their velocity will cancel each other's out transforming this energy into heat. While the main mass of each star will continue moving as they did before the collision. The third body created from the two sheared parts of the stars will form between the two original stars.

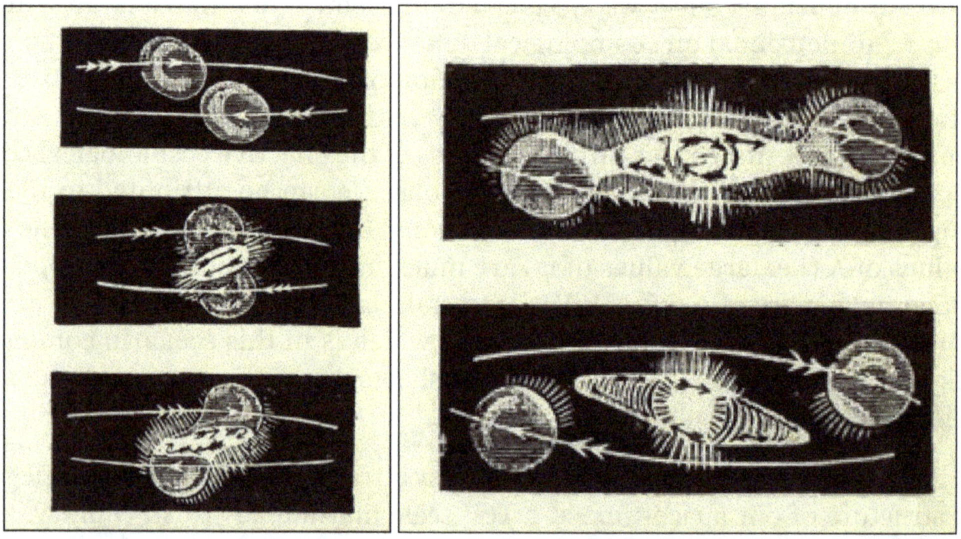

Diagram from The Romance of the Heavens explaining partial impact theory.

The temporary star expands after the impact displaying an intense increase in light; after all molecular reactions have taken place the light is replaced by a hollow shell of gas or possibly a planetary nebula, and eventually dissipates into space. Bickerton explains this bright temporary star by saying that it doesn't disappear due to cooling, but that it was too hot to hold together.

The temperature of the third star isn't dependent on the amount of contact between the two original stars, but rather the chemical makeup of the stars and their velocities going into the collision. The stability of the third body depends on the size of the contact of the original stars, if the contact was small then the mass of the newly created third body will find it harder to attract molecules to it. Rather than if it had a larger mass where molecules would find it more difficult to escape from its larger gravitational pull.

Rainbow Gravity Theory

After Einstein proposed the general theory of relativity describing the gravitational force in terms of space time curvatures, the proposed field equations related the space time geometry to the matter content of the universe.

In general relativity, the universe is modeled as a space time, which has mathematically a structure of a four dimensional differentiable manifold. This means that locally the space time is always flat, in a sufficiently small region around any point, but on a larger scale it does not have to be so and it can have more rich and varied structure. A two-dimensional example of such a manifold is a sphere, which is flat enough in the vicinity of any single point on its surface, but has a non-zero global curvature.

The earliest solutions found for the field equations were the Schwarzschild metric representing the gravitational field around an isolated body such as a spherically symmetric star, and the Friedmann cosmological models. Both these contained a space time singularity where the curvatures and energy densities were infinite and the physical description would then break down. In the Schwarzschild solution such a singularity was present at the center of symmetry r = 0 whereas for the Friedmann models it is found at the epoch t = 0 which is beginning of the universe and origin of time where the scale factor for the universe vanishes and all objects are crushed to a zero volume due to infinite gravitational tidal forces.

Even though the physical problem posed by the existence of such a strong curvature singularity was realized immediately in these solutions, which turned out to have several important implications towards the experimental verification of the general relativity theory, initially this phenomenon was not taken seriously. It was generally thought that the existence of such a singularity must be a consequence of the very high degree of symmetry imposed on the space time while deriving these solutions. Subsequently, the distinction between a genuine space time singularity and a mere coordinate singularity became clear and it was realized that the singularity at r = 2m in the Schwarzschild space time was only a coordinate singularity which could be removed by a suitable coordinate transformation. It was clear, however, that the genuine curvature singularity at r = 0 cannot be removed by any such transformations. The hope was then that when more general solutions are considered with a less degree of symmetry requirements, such singularities will be avoided. This issue was sorted out when a detailed study of the structure of a general space time and the associated problem of singularities was taken up by Hawking, Penrose, and Geroch, which showed that singularities are in fact much more general phenomena in gravitation theories.

Further to the general relativity theory in 1915, the gravitation physics was a relatively quiet field with few developments till about 1950s. However, the 1960s saw the emergence of new observations in high energy astrophysics, such as quasars and high energy phenomena at the center of galaxies such as extremely energetic jets. These observations, together with important theoretical developments such as studying the global structure of space times and singularities, led to important results in black hole physics and relativistic astrophysics and cosmology.

Space Time Singularities

The work in early 1970s in gravitation theories showed that a space time will admit singularities within a rather general framework provided it satisfies certain reasonable physical assumptions such as the positivity of energy, a suitable causality assumption and a condition implying strong gravitational fields, such as the existence of trapped surfaces. It thus followed that singularities form a rather general feature of the relativity theory. In fact, these considerations ensure the existence of singularities in other theories of gravity also which are based on a space time manifold framework and that satisfy the general conditions.

Therefore the scenario that emerges is, essentially for all classical space time theories of gravitation, the occurrence of singularities form an inevitable and integral part of the description of the physical reality. In the vicinity of such a singularity, typically the energy densities, space time curvatures, and all other physical quantities would blow up, thus indicating the occurrence of super ultra-dense regions in the universe. The behaviour of such regions may not be governed by the classical theory itself, which may breakdown having predicted the existence of the singularities, and a quantum gravitational theory would be the likely description of the phenomena created by such singularities.

Firstly, it is to be clarified how to characterize a space time singularity. It turns out that it is the notion of geodesic incompleteness that characterizes a singularity in an effective manner for a space time and enables their existence to be proved by means of general enough theorems, which involve a consideration of the gravitational focusing caused by the space time curvature in congruence of time like and null geodesics. This turns out to be the main cause of the existence of singularity in the form of non-space like incomplete geodesics in space time. The issue of physical nature of a space time singularity is important. There are many types of singular behaviours possible for a space time and some of these could be regarded as mathematical pathologies in the space time rather than having any physical significance. This will be especially so if the space time curvature and similar other physical quantities remained finite along an incomplete non space like geodesic in the limit of approach to the singularity. A singularity will be physically important when there is a powerful enough curvature growth along singular geodesics, and the physical interpretation and implications of the same are to be considered.

Considering various situations, the occurrence of non-space like geodesic incompleteness has been generally agreed upon as the criterion for the existence of a singularity for a space time. It is clear that if a space time manifold contains incomplete non-space like geodesics, there is a definite singular behaviour present in the space time. In such a case, a time like observer or a photon suddenly disappears from the space time after a finite amount of proper time or after a finite value of the affine parameter. The singularity theorems which result from an analysis of gravitational focusing and global properties of a space time prove this incompleteness property for a wide class of space times under a set of rather general conditions.

The matter fields with positive energy density affect the causality relations in a space time and cause focusing in the families of time like and null trajectories. The essential phenomena that occurs here is that matter focuses the non-space like geodesics of the space time into pairs of focal points or the conjugate points. The rate of change of volume expansion for a given congruence of time like geodesics can be written as:

$$\frac{d\theta}{d\tau} = -R_{lk}V^l V^k - \frac{1}{3}\theta^2 - 2\sigma^2 + 2w^2$$

where, for a given congruence of time like geodesics, the quantities θ, σ and ω are expansion, shear, and rotation tensors are respectively. The above equation is called the Raychaudhuri equation which describes the rate of change of the volume expansion as one move along the time like geodesic curves in the congruence. We note that the second and third term on the right-hand side involving θ and σ are positive always. Consider now the term $R_{ij}V^iV^j$. By Einstein equations this can be written as:

$$R_{ij}V^iV^j = 8\pi\left[T_{ij}V^iV^j + \frac{1}{2}T\right]$$

The term $T_{ij}V^iV^j$ above represents the energy density as measured by a time like observer with the unit tangent V^i, which is the four-velocity of the observer. For all reasonable classical physical fields this energy density is generally taken as non-negative and it is assumed that for all time like vectors V^i the following is satisfied.

$$T_{ij}V^iV^j \geq 0$$

Such an assumption is called the weak energy condition. When a suitable energy condition is satisfied, the Raychaudhuri equation implies that the effect of matter on space time curvature causes a focusing effect in the congruence of time like geodesics due to gravitational attraction. This, in general causes the neighboring geodesics in the congruence to cross each other to give rise to caustics or conjugate points. This separation between nearby time like geodesics is governed by what is called the geodesic deviation equation,

$$D^2Z^j = -R^j{}_{kil}V^kZ^iV^l$$

where Z^i is the separation vector between nearby geodesics of the congruence. Solutions of the above equation are called the Jacobi fields along a given time like geodesic. There are several singularity theorems available which establish the non-space like geodesic incompleteness for a space time under different sets of conditions and applicable to different physical situations. However, the most general of these is the Hawking–Penrose theorem, which is applicable in both the collapse situation and cosmological scenario. The main idea of the proof of such a theorem is, using the causal structure analysis it is shown that there must be maximal length time like curves between certain pairs of events in the space time. Now, a causal geodesic which is both future and past complete must contain pairs of conjugate points if M satisfies the generic condition and an energy condition.

This is then used to draw the necessary contradiction to show that M must be non-space like geodetically incomplete. The inevitable existence of space time singularities, for wide classes of general models of space times means that the classical gravity necessarily gives rise to regions in the space time universe where the densities and space time curvatures would really grow without any bounds, where all other physical parameters also would diverge really.

Gravitational Collapse

The existence of space time singularities in the Einstein gravity, and in all similar space time theories of gravitation poses intriguing challenges and fundamental questions in physics as well as cosmology. Such a phenomenon will basically arise in two physical scenarios in the universe, the first being the cosmology where such a singularity will correspond to the origin of the universe, and secondly whenever locally a large quantity of matter and energy collapses under the force of its own gravity. This later situation will be effectively realized in the gravitational collapse of massive stars in the universe, which collapse and shrink catastrophically under their self-gravity, when the star has exhausted its nuclear fuel within which earlier supplied the internal pressure to halt the in-fall due to gravity.

When a massive star, more than a few solar masses, has exhausted its internal nuclear fuel, it is believed to enter the stage of an endless gravitational collapse without having any final equilibrium state. According to the Einstein theory of gravitation, the star goes on shrinking in its radius, reaching higher and higher densities. What would be the final fate of such an object according to the general theory of relativity? This is one of the central questions in relativistic astrophysics and gravitation theory today. It has been suggested that the ultra-dense object that forms as a result of collapse could be a black hole in the space and time from which not even light rays can escape. Alternatively, if the event horizon of gravity fails to cover the final crunch, it could be a visible singularity which can causally interact with the outside universe and from which emissions of light and matter may be possible.

It is of course reasonably clear that very near such a space time singularity, the classical description that predicted it must itself breakdown. The quantum effects associated with gravity are most likely to become dominant in such a regime. These may resolve the classical singularity. However, we have no viable and consistent quantum theory of gravity available as of today despite many serious attempts, and therefore the issue of resolution of singularities as produced by classical gravity remains very much open currently.

An investigation on final fate of collapse is of importance from both the theoretical as well as observational point of view. At the theoretical level, working out the collapse outcomes in general relativity is crucial to the problem of asymptotic predictability, namely, whether the singularities forming at the end point of collapse will be necessarily covered by the event horizons of gravity. A hypothesis that remains fundamental to the theoretical foundations of black hole physics and its numerous astrophysical applications which have been invoked in past decades (e.g. the area theorem for black holes, laws of black hole thermodynamics, Hawking radiation effect, predictability; and on observational side, accretion of matter by black holes, massive black holes at the center of galaxies etc.), is cosmic censorship which states the singularities of collapse must be hidden within horizons of gravity. On the other hand, existence of visible or naked singularities would offer a new approach on these issues requiring modification and reformulation of our usual theoretical conception on black holes.

To investigate this issue, dynamical collapse scenarios have been examined in past decade or so for many cases such as clouds composed of dust, radiation, perfect fluids, or also of matter compositions with more general equations of state.

Black Holes

One could consider a gravitationally collapsing spherical massive star. We need to consider the interior solution for the object which will depend on the properties of matter, equation of state, and the physical processes taking place within the stellar interior. However, assuming the matter to be pressure less dust allows solving the problem analytically, providing many important insights. Here the energy-momentum tensor is given by $T^{ij} = \rho u^i u^j$, and one needs to solve the Einstein equations for the spherically symmetric metric. This determines the metric potentials, and the interior geometry of the collapsing dust ball is given by,

$$ds^2 = -dt^2 + R^2(t)\left[\frac{dr^2}{1-r^2} + r^2 d\Omega^2\right]$$

where $d\Omega^2 = d\theta^2 + \sin^2\theta d\varphi^2$ is the metric on two-sphere. The geometry outside is vacuum

Schwarzschild space-time. The interior geometry of the dust cloud matches at the boundary $r = r_b$ with the exterior Schwarzschild space-time.

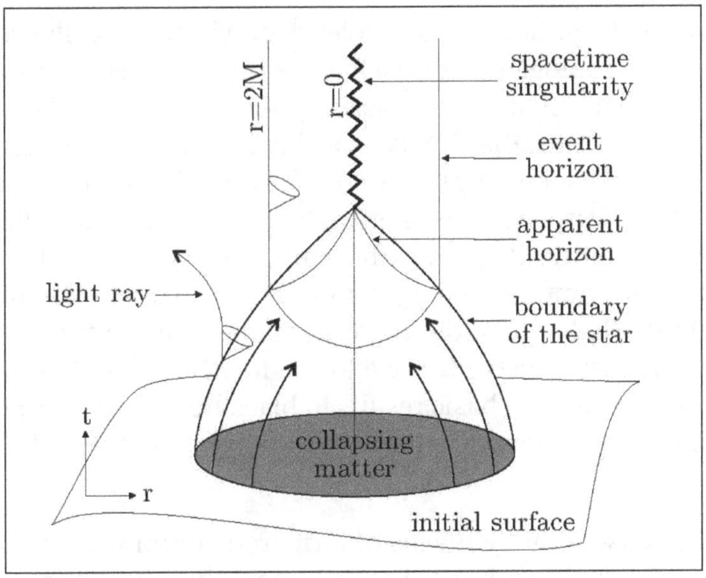

Spherically symmetric homogeneous dust collapse where the final outcome is black hole formation in the space time.

The basic features of such a collapsing, spherical, homogeneous dust cloud configuration are given in Fig. The collapse is initiated when the star surface is outside its Schwarzschild radius r = 2m, and a light ray emitted from the surface of the star can escape to infinity. However, once the star has collapsed below r = 2m, a black hole region of no escape develops in the space-time, bounded by the event horizon at r = 2m. Any point in this empty region represents a trapped surface (which is a two-sphere in space-time) in that both the outgoing and ingoing families of null geodesics emitted from this point converge and hence no light comes out of this region. Then, the collapse to an infinite density and curvature singularity at r = 0 becomes inevitable in a finite proper time as measured by an observer on the surface of the star. The black hole region in the resulting vacuum Schwarzschild geometry is given by 0 < r < 2m, the event horizon being the outer boundary. On the event horizon, the radial outwards photons stay where they are, but all the rest are dragged inwards. No information from this black hole can propagate outside r = 2m to observers far away. We thus see that the collapse gives rise to a black hole in the space-time which covers the resulting space-time singularity. The ultimate fate of the star undergoing such a collapse is then an infinite curvature singularity at r = 0, which is completely hidden within the black hole. No emissions or light rays from the singularity could go out to observer at infinity and the singularity is causally disconnected from the outside space-time.

Cosmic Censorship

The question now is whether one could generalize these conclusions on the occurrence of a space time singularity in collapse and black hole formation for more general matter fields, and for non-spherical situations.

While we know that the occurrence of the singularity itself is stable to small perturbations in the initial data, there is no proof available that such a singularity will continue to be hidden within a black hole and remain causally disconnected from outside observers, even when the collapse is not spherical or when the matter does not have the form of exact homogeneous dust.

Thus, in order to generalize the notion of black holes to gravitational collapse situations other than exact spherically symmetric homogeneous dust case, it becomes necessary to rule out such naked or visible singularities by means of an explicit assumption. This is stated as the cosmic censorship hypothesis, which essentially states that if S is a partial Cauchy surface from which collapse commences, then there are no naked singularities to the future of S, that is, which could be seen from the future null infinity. This is true for the spherical homogeneous dust collapse, where the resulting space time is future asymptotically predictable and the censorship holds. Thus, the breakdown of physical theory at the space time singularity does not disturb prediction in future for the outside asymptotically flat region. What will be the corresponding scenario for other collapse situations, when in homogeneities, non-sphericity etc are allowed for? It is clear that the assumption of censorship in a suitable form is crucial to basic results in black hole physics. In fact, when one considers the gravitational collapse in a generic situation, the very existence of black holes requires this hypothesis.

If one is to establish the censorship by means of a rigorous proof that of course requires a much more precise formulation of the hypothesis. The statement that results of a complete gravitational collapse must always be a black hole and not a naked singularity, or all singularities of collapse must be hidden in black holes, causally disconnected from observers at infinity, is not rigorous enough. This is because, under completely general circumstances, the censorship or asymptotic predictability is false as one could always choose a space-time manifold with a naked singularity which would be a solution to Einstein's equations if we define $T_{ij} \equiv (1/8\pi)G_{ij}$ In fact, as far as the cosmic censorship hypothesis is concerned, it is a major problem in itself to find a satisfactory and mathematically rigorous formulation of what is physically desired to be achieved. Developing a suitable formulation would probably be a major advance towards the solution of the main problem. It should be noted that presently no general proof is available for any suitably formulated version of the weak censorship.

Naked Singularities

This situation leads us to conclude that the first and foremost task is to carry out a detailed and careful examination of various gravitational collapse scenarios to examine them for their end states. It is clear from these considerations that we still do not have sufficient data and information available on the various possibilities for gravitationally collapsing configurations so as to decide one way or other on the issue of censorship hypothesis. What appears really necessary is a detailed investigation of different collapse scenarios, and to examine the possibilities arising, in order to have insights into the issue of the final fate of gravitational collapse. With such a purpose, several gravitational collapse scenarios involving different forms of matter have been investigated to understand better the final fate of collapse.

Since we are interested in collapse, we require that the space-time contains a regular initial space like hyper surface on which the matter fields, as represented by the stress-energy tensor T_{ij}, have a compact support and all physical quantities are well-behaved on this surface. Also, the matter should satisfy a suitable energy condition and the Einstein equations are satisfied. We say that the space-time contains a naked singularity if there is a future directed non-space like curve which reaches a faraway observer or infinity in future, and in the past it terminates at the singularity.

As an immediate generalization of the Oppenheimer-Snyder-Datt homogeneous dust collapse, one could consider the collapse of inhomogeneous dust and examine the nature and structure of resulting singularity with special reference to censorship, and the occurrence of black holes and naked singularities. The main motivation to discuss this situation is this provides a clear picture in an explicit manner of what is possible in gravitational collapse. One could ask how are the conclusions given above for homogeneous collapse are modified when the in homogeneities of matter distribution are taken into account. Clearly, it is important to include effects of in homogeneities because typically a realistic collapse would start from a very inhomogeneous initial data with a centrally peaked density profile.

This problem was investigated in detail using the Tolman-Bondi-Lemaitre models, which describe gravitational collapse of an inhomogeneous spherically symmetric dust cloud. This is an infinite dimensional family of asymptotically flat solutions of Einstein equations, which is matched to the Schwarzschild space time outside the boundary of the collapsing star. The Oppenheimer-Snyder-Datt model is a special case of this class of solutions.

It is seen that the introduction of in homogeneities leads to a rather different picture of gravitational collapse. The metric for spherically symmetric collapse of inhomogeneous dust, in commoving coordinates (t, r, θ, φ), is given by,

$$ds^2 = -dt^2 + \frac{R'^2}{1+f} dr^2 + R^2\left(d\theta^2 + \sin^2\theta d\varphi^2\right)$$

$$T^{ij} = \varepsilon\delta^i_t\delta^i_t\,,\ \varepsilon = \varepsilon(t,r) = \frac{F'}{R^2 R'}$$

where T^{ij} is the stress-energy tensor, ε is the energy density, and R is a function of both t and r given by,

$$\dot{R}^2 = \frac{F}{R} + f$$

Here the dot and prime denote partial derivatives, with respect to the parameters t and r respectively. As we are considering collapse, we require $R(t,r) < 0$. The quantities F and f are arbitrary functions of r and $4\pi R^2$ (t, r) is the proper area of the mass shells. The area of such a shell at r = const. goes to zero when R (t, r) = 0. For gravitational collapse situation, we take ϱ to have compact support on an initial space like hyper surface and the space-time can be matched at some r = const. = r_c to the exterior Schwarzschild field with total Schwarzschild mass m (r_c) = M enclosed within the dust ball of coordinate radius of r = r_c. The apparent horizon in the interior dust ball lies at R = F(r).

Using this framework, the nature of the singularity R = 0 can be examined. In particular, the problem of nakedness or otherwise of the singularity can be reduced to the existence of real, positive roots of an algebraic equation V (X) = 0, constructed out of the free functions F and f and their derivatives which constitute the initial data of this problem. If the equation V (X) = 0 has a real positive root, the singularity could be naked. In order to be the end point of null geodesics at least one real positive value of X_0 should satisfy the above. Clearly, if no real positive root of the above is found, the singularity is not naked. It should be noted that many real positive roots of the above equation may exist which give the possible values of tangents to the singular null geodesics terminating at the singularity. Suppose now X = Xo is a simple root to V (X) = 0. To determine whether Xo is realized as a tangent along any outgoing singular geodesics to give a naked singularity, one can integrate the equation of the radial null geodesics in the form r = r(X) and it is seen that there is always at least one null geodesic terminating at the singularity t = 0, r = 0, with $X = X_0$. In addition there would be infinitely many integral curves as well, depending on the values of the parameters involved that terminate at the singularity. It is thus seen that the existence of a positive real root of the equation V (X) = 0 is a necessary and sufficient condition for the singularity to be naked. Finally, to determine the curvature strength of the naked singularity at t = 0, r = 0, one may analyze the quantity $k^2R_{ab}K^aK^b$ near the singularity. Standard analysis shows that the strong curvature condition is satisfied, in that the above quantity remains finite in the limit of approach to the singularity.

General Collapse Scenarios

The assumption of vanishing pressures, which could be important in the final stages of the collapse, may be considered as the limitation of dust models. On the other hand, it is also argued sometimes that in the final stages of collapse, the dust equation of state could be relevant and at higher and higher densities the matter may behave more like dust. Further, if there are no large negative pressures (as implied by the validity of the energy conditions), then the pressure also might contribute gravitationally in a positive manner to the effect of dust and may not alter the conclusions.

In any case, it is important to consider collapse situations with non-zero pressures and with reasonable equations of state. Pressures may play an important role for the later stages of collapse and one must investigate the possibility if pressure gradients could prevent the occurrence of naked singularity. These issues have been examined namely, the existence, the termination of non-space like geodesic families, and the strength of such a singularity for collapse with non-zero pressure. The results could be summarized as follows. If in a self-similar collapse with pressure, a single null radial geodesic escapes the singularity, then an entire family of non-space like geodesics would also escape provided the positivity of energy density is satisfied as above.

Actually, gravitational collapse models with a general form of matter, together with those such as directed radiation, dust, perfect fluids etc. imply some general pattern emerging about the final outcome of gravitational collapse. Basically it follows that the occurrence of naked singularity is basically related to the choice of initial data to the Einstein field equations, and would therefore occur from regular initial data within the general context considered, subject to the matter satisfying weak energy condition. It appears that the occurrence of naked singularity or a black hole is more a problem of choice of the initial data for field equations rather than that of the form of matter or the equation of state. This has important implication for cosmic censorship in that in order to preserve the same one has to avoid all such regular initial data causing naked singularity, and hence a deeper understanding of the initial data space is required in order to determine such initial data and the kind of physical parameters they would specify. This would, in other words, classify the range of physical parameters to be avoided for a particular form of matter. More importantly, it would also pave the way for the black hole physics to use only those ranges of allowed parameter values which would produce black holes, thus putting black hole physics on a more firm footing.

What will be the final fate of gravitational collapse which is not spherically symmetric? The main phases of spherical collapse of a massive star would be typically instability, implosion of matter, and subsequent formation of an event horizon and a space-time singularity of infinite density and curvature with infinite gravitational tidal forces. Again, small perturbations over the spherically symmetric situation would leave the situation unchanged in the sense that an event horizon will continue to form in the advanced stages of the collapse.

The question then is, do horizons still form when the fluctuations from the spherical symmetry are high and the collapse is highly non-spherical? It was shown by Thorne, that when there is no spherical symmetry, the collapse of infinite cylinders do give rise to naked singularities in general relativity, which are not covered by horizons. This situation motivated Thorne to propose the following hoop conjecture for finite systems in an asymptotically flat space-time, which characterizes the final fate of non-spherical collapse: The horizons of gravity form when and only when a mass M gets compacted in a region whose circumference in every direction obeys $C \leq 2\pi(2GM/c^2)$. Thus, unlike the cosmic censorship conjecture, the hoop conjecture does not rule out all the naked singularities but only makes a definite assertion on the occurrence of the event horizons in gravitational collapse. We also note that the hoop conjecture is concerned with the formation of event horizons, and not with naked singularities. Thus, even when event horizons form, say for example in the spherically symmetric case, it does not rule out the existence of naked singularities, i.e. it does not imply that such horizons must always cover the singularities.

Apart from such numerical simulations, some analytic treatments of aspherical collapse are also available. For example, the aspherical Szekeres models for irrotational dust without any Killing vectors, generalizing the spherical Tolman-Bondi-Lemaitre collapse, were studied, to deduce the existence of strong curvature central naked singularities. While this indicates that naked singularities are not necessarily confined to spherical symmetry, it must be noted that dynamical evolution of a non-spherical collapse still remains a largely uncharted territory.

We note here that the genericity and stability of the collapse outcomes, in terms of black holes and naked singularities need to be understood carefully and in further detail. It is by and large well-accepted now, that the general theory of relativity does allow and gives rise to both black holes and naked singularities as final fate of a continual gravitational collapse, evolving from a regular initial data, and under reasonable physical conditions. What is not fully clear as yet is the distribution of these outcomes in the space of all allowed outcomes of collapse. The collapse models and considerations we gave here would be of some help in this direction, and may throw some light on the distribution of black holes and naked singularity solutions in the initial data space.

The important question then is the genericity and stability of such naked singularities arising from regular initial data. Will the initial data subspace, which gives rise to naked singularity as end state of collapse, have zero measure in a suitable sense? In that case, one would be able to reformulate more suitably the censorship hypothesis, based on a criterion that naked singularities could form in collapse but may not be generic.

It is natural to ask here, what is really the physics that causes a naked singularity to develop in collapse, rather than a black hole? We need to know how at all particles and energy are allowed to escape from extremely strong gravity fields. We have examined this issue in some detail to bring out the role of in homogeneities and space-time shear to achieve this towards distorting the geometry of horizons forming in collapse. In Newtonian gravity, it is only the matter density that determines the gravitational field. In Einstein theory, however, density is just one attribute of the overall gravitational field, and the various curvature components and scalar quantities play an equally important role to dictate what the overall nature of the field is. Once the density is inhomogeneous or higher at the center of collapsing star, this rather naturally delays the trapping of light and matter during collapse, which can in principle escape away. This is a general relativistic effect wherein even if the densities are very high, paths are created for light or matter to escape due to in homogeneously collapsing matter fields, and these physical features naturally lead to a naked singularity formation rather than a black hole end state. It is the amount of inhomogeneity that counts to distort the horizons. If it is very small, below a critical limit, a black hole will form, but with sufficient inhomogeneity trapping is delayed to cause a naked singularity. This criticality again comes out in the Vaidya class of radiation collapse models, where it is the rate of collapse, that is how fast or slow the cloud is collapsing, that determines the black hole or naked singularity formation.

Distinguishing Black Holes and Naked Singularities Observationally

It is clear that the black hole and naked singularity outcomes of a complete gravitational collapse for a massive star are very different from each other physically, and would have quite different observational signatures. In the naked singularity case, if it occurs in nature, we have the possibility to observe the physical effects happening in the vicinity of the ultra-dense regions that form in the very final stages of collapse. However, in a black hole scenario, such regions are necessarily hidden within the event horizon of gravity.

There have been attempts where researchers explored physical applications and implications of the naked singularities. If we could find astrophysical applications of the models that predict naked singularities as collapse final fate, and possibly try to test the same through observational methods and the signatures predicted, that could offer a very interesting avenue to get further insight into the problem as a whole. An attractive recent possibility in that connection is to explore the naked singularities as possible particle accelerators.

Also, the accretion discs around a naked singularity, wherein the matter particles are attracted towards or repulsed away from the singularities with great velocities could provide an excellent venue to test such effects and may lead to predictions of important observational signatures to distinguish the black holes and naked singularities in astrophysical phenomena. The question of what observational signatures would then emerge and distinguish the black holes from naked singularities is then necessary to be investigated, and we must explore what special astrophysical consequences the latter may have.

Where could the observational signatures of naked singularities lie? If we look for the sign of singularities such as the ones that appear at the end of collapse, we have to consider explosive and high energy events. In fact such models expose the ultra-high density region at the time of formation of the singularity while the outer shells are still falling towards the center. In such a case, shockwaves emanating from the super dense region at scales smaller than the Schwarzschild radius (that could be due to quantum effects or repulsive classical effects) and collisions of particles near the Cauchy horizon could have effects on the outer layers. These would be considerably different from those appearing during the formation of a black hole. If, on the other hand, we consider singularities such as the super spinning Kerr solution we can look for different kinds of observational signatures. Among these the most prominent features deal with the way the singularity could affect incoming particles, either in the form of light bending, such as in gravitational lensing, particle collisions close to the singularity, or properties of accretion disks.

Essentially we ask whether we could test censorship using astronomical observations. With so many high technology power missions to observe the cosmos, can we not just observe the skies carefully to determine the validity or otherwise of the cosmic censorship? In this connection, several proposals to measure the mass and spin ratio for compact objects and for the galactic center have been made by different researchers. In particular, using pulsar observations it is suggested that gravitational waves and the spectra of X-rays binaries could test the rotation parameter for the center of our galaxy. Also, the shadow cast by the compact object can be used to test the same in stellar mass objects, or X-ray energy spectrum emitted by the accretion disk can be used. Using certain observable properties of gravitational lensing that depend upon rotation is also suggested.

The basic issue here is that of sensitivity, namely how accurately and precisely can we measure and determine these parameters. A number of present and future astronomical missions could be of help. One of these is the Square-Kilometer Array (SKA) radio telescope, which will offer a possibility here, with a collecting area exceeding a factor of hundred compared to existing ones. The SKA astronomers point out they will have the sensitivity desired to measure the required quantities very precisely to determine the vital fundamental issues in gravitation physics such as the cosmic censorship, and to decide on its validity or otherwise. Other missions that could in principle provide a huge amount of observational data are those that are currently hunting for the gravitational waves. Gravitational wave astronomy has yet to claim its first detection of waves, nevertheless in the coming years it is very likely that the first observations will be made by the experiments such as LIGO and VIRGO that are currently still below the threshold for observation. Then gravitational wave astronomy will become an active field with possibly large amounts of data to be checked against theoretical predictions and it appears almost certain that this will have a strong impact on open theoretical issues such as the Cosmic Censorship problem.

There are three different kinds of observations that one could devise in order to distinguish a naked singularity from a black hole. The first one relies on the study of accretion disks. The accretion properties of particles falling onto a naked singularity would be very different from those of black hole of the same mass and the resulting accretion disks would also be observationally different. The properties of accretion disks have been studied in terms of the radiant energy, flux and luminosity, in a Kerr-like geometry with a naked singularity, and the differences from a black hole accretion disk have been investigated. Also, the presence of a naked singularity gives rise to powerful repulsive forces that create an outflow of particles from the accretion disk on the equatorial plane. This outflow that is otherwise not present in the black hole case could be in principle distinguished from the jets of particles that are thought to be ejected from black hole's polar region and which are due to strong electromagnetic fields. Also, when charged test particles are considered the accretion disk's properties for the naked singularity present in the Reissner-Nordstrom space time are seen to be observationally different from those of black holes.

The second way of distinguishing black holes from naked singularities relies on gravitational lensing. It is argued that when the space time does not possess a photon sphere, then the lensing features of light passing close to the singularity will be observationally different from those of a black hole. This method, however, does not appear to be very effective when a photon sphere is present in the space time. Assuming that a Kerr-like solution of Einstein equations with massless scalar field exists at the center of galaxies, its lensing properties are studied and it was found that there are effects due to the presence of both the rotation and scalar field that would affect the behavior of the bending angle of the light ray, thus making those objects observationally different from black holes.

Finally, a third way of distinguishing black holes from naked singularities comes from particle collisions and particle acceleration in the vicinity of the singularity. In fact, it is possible that the repulsive effects due to the singularity can deviate a class of infalling particles, making these outgoing eventually. These could then collide with some ingoing particle, and the energy of collision could be arbitrarily high, depending on the impact parameter of the outgoing particle with respect to the singularity. The net effect is thus creating a very high energy collision that resembles that of an immense particle accelerator and that would be impossible in the vicinity of a Kerr black hole.

Predictability and other Cosmic Puzzles

What then is the status of naked singularities versus censorship today? Can cosmic censorship survive in some limited and specialized form, and firstly, can we properly formulate it after all these studies in recent years on gravitational collapse? While this continues to be a major cosmic puzzle, recent studies on formation of naked singularities as collapse end states for many realistic models have brought to forefront some of the most intriguing basic questions, both at classical and quantum level, which may have significant physical relevance. Some of these are: Can the super ultra-dense regions forming in a physically realistic collapse of a massive star be visible to far away observers in space-time? Are there any observable astrophysical consequences? What is the causal structure of space-time in the vicinity of singularity as decided by the internal dynamics of collapse which evolves from a regular initial data at an initial time? How early or late the horizons will actually develop in a physically realistic gravitational collapse, as determined by the astrophysical conditions within the star? When a naked singularity forms, is it possible to observe the quantum gravity effects taking place in the ultra-strong gravity regions? Can one possibly envisage a connection to observed ultra-high energy phenomena such as cosmic gamma ray bursts?

A continuing study of collapse phenomena within a general and physically realistic framework may be the only way to answers on some of these issues. This could lead us to novel physical insights and possibilities emerging out of the intricacies of gravitational force and nature of gravity, as emerging from examining the dynamical evolutions as allowed by Einstein equations. Apart from its physical relevance, the collapse phenomena also have profound philosophical implications such as on the issue of predictability in the universe.

It is sometimes argued that breakdown of censorship means violation of predictability in space time, because we have no direct handle to know what a naked singularity may radiate and emit unless we study the physics in such ultra-dense regions. One would not be able then to predict the universe in the future of a given epoch of time as would be the case, for example, in the case of the Schwarzschild black hole that develops in Oppenheimer-Snyder collapse.

A concern usually expressed is that if naked singularities occurred as the final fate of gravitational collapse that would break the predictability in the space time, because the naked singularity is characterized by the existence of light rays and particles that emerge from the same. Typically, in all the collapse models discussed above, there is a family of future directed non-space like curves that reach external observers, and when extended in the past these meet the singularity. The first light ray that comes out from the singularity marks the boundary of the region that can be predicted from a regular initial Cauchy surface in the space time, and that is called the Cauchy horizon for the space time. The causal structure of the space time would differ significantly in the two cases, when there is a Cauchy horizon and when there is none.

In general relativity, a given 'epoch' of time is sometimes represented by a space like surface, which is three-dimensional space. For example, in the standard Friedmann models of cosmology, there is such an epoch of simultaneity, from which the universe evolves in future, given the physical variables and initial data on this surface. The Einstein equations govern this evolution of universe, and there is thus a predictability which one would expect to hold in a classical theory. The concern that is expressed at times is one would not be able to predict in the future of naked singularity, and that unpredictable inputs may emerge from the same.

The point here is, given a regular initial data on a space like hyper surface, one would like to predict the future and past evolutions in the space time for all times. Such a requirement is termed as the global hyperbolicity of the space time. A globally hyperbolic space time is a fully predictable universe, it admits a Cauchy surface, which is a three dimensional space like surface the data on which can be evolved for all times in the past as well as in future. Simple enough space times such as the Minkowski or Schwarzschild are globally hyperbolic, but the Reissner-Nordstrom or Kerr geometries are not globally hyperbolic.

The key role that the event horizon of a black hole plays is that it hides the super-ultra-dense region formed in collapse from us. So the fact that we do not understand such regions has no effect on our ability to predict what happens in the universe at large. But if no such horizon exists, then the ultra-dense region might, in fact, play an important and even decisive role in the rest of the universe, and our ignorance of such regions would become of more than merely academic interest.

Yet such unpredictability is common in general relativity, and not always directly related to censorship violation. Even black holes themselves need not fully respect predictability when they rotate or have some charge. For example, if we drop an electric charge into an uncharged black hole, the space time geometry radically changes and is no longer predictable from a regular initial epoch of time. A charged black hole admits a naked singularity which is visible to an observer within the horizon, and similar situation holds when the black hole is rotating. There is an important debate in recent years, if one could over-charge or over-rotate a black hole so that the singularity visible to observers within the horizon becomes visible to external far away observers too.

Another point is, if such a black hole was big enough on a cosmological scale, the observer within the horizon could survive in principle for millions of years happily without actually falling into the singularity, and would thus be able to observe the naked singularity for a long time. Thus, only purest of pure black holes with no charge or rotation at all respect the full predictability, and all other physically realistic ones with charge or rotation actually do not. As such, there are many models of the universe in cosmology and relativity that are not totally predictable from a given space like hyper surface in the past. In these universes, the space time cannot be neatly separated into space and time foliation so as to allow initial data at a given moment of time to fully determine the future.

The real breakdown of predictability is the occurrence of space time singularity itself, which indicates the true limitation of the classical gravity theory. It does not matter really whether it is hidden within an event horizon or not. The real solution of the problem would then be the resolution of singularity itself, through either a quantum theory of gravity or in some way at the classical level itself.

Actually, the cosmic censorship way to predictability, that of 'hiding the singularity within a black hole', and then thinking that we restored the space time predictability may not be the real solution, or at best it may be only a partial solution to the key issue of predictability in space time universes. In fact, it may be just shifting the problem elsewhere, and some of the current major paradoxes faced by the black hole physics such as the information paradox, the various puzzles regarding the nature of the Hawking radiation, and other issues could as well be a manifestation of the same.

No doubt, the biggest argument in support of censorship would be that it would justify and validate the extensive formalism and laws of black hole physics and its astrophysical applications made so far. Censorship has been the foundation for the laws of black holes such as the area theorem and others, and their astrophysical applications. But these are not free of major paradoxes. Even if we accept that all massive stars would necessarily turn into black holes, this still creates some major physical paradoxes. Firstly, all the matter entering a black hole must of necessity collapse into a space-time singularity of infinite density and curvatures, where all known laws of physics break down, which is some kind of instability at the classical level itself. This was a reason why many gravitation theorists of 1940s and 1950s objected to black hole formation, and Einstein also repeatedly argued against such a final fate of a collapsing star. Also, as is well-known and has been widely discussed in the past few years, a black hole, by potentially destroying information, appears to contradict the basic principles of quantum theory. In that sense, the very formation of a black hole itself with a singularity within it appears to come laden with inherent problems. It is far from clear how one would resolve these basic troubles even if censorship were correct.

In view of such problems with the black hole paradigm, a possibility worth considering is the delay or avoidance of horizon formation as the star collapses under gravity. This happens when collapse to a naked singularity takes place, namely, where the horizon does not form early enough or is avoided. In such a case, if the star could radiate away most of its mass in the late stages of collapse, this may offer a way out of the black hole conundrum, while also resolving the singularity issue, because now there is no mass left to form the singularity.

What this means is, such'unpredictability' is somewhat common in general relativity. For example, if we drop a slight charge in a Schwarzschild black hole, the space time geometry completely changes into that of a charged black hole that is no longer predictable in the above sense. Similar situation holds when the black hole is rotating. In fact, there are very many models of universe in use in relativity which are not 'globally hyperbolic', that is, not totally predictable in the above sense where space and time are neatly separated so as to allow initial data to fully determine future for all times.

In any case, a positive and useful feature that has emerged from work on collapse models so far is, we already have now several important constraints for any possible formulation of censorship. It is seen that several versions of censorship proposed earlier would not hold, because explicit counter-examples are available now. Clearly, analyzing gravitational collapse plays a crucial role here. Only if we understand clearly why naked singularities do develop as collapse end states in many realistic models, there could emerge any pointer or lead to any practical and provable version of censorship.

Finally, it may be worth noting that even if the problem of singularity was resolved somehow, possibly by invoking quantum gravity which may smear the singularity, we still have to mathematically formulate and prove the black hole formation assuming an appropriate censorship principle, which is turning out to be most difficult task with no sign of resolve.

4

Astrophysics of Galaxies

Galaxies are the vast cosmic islands of stars, gases, dust and dark matter that span the vast universe. All aspects related to galaxies are addressed under the astrophysics of galaxies. The formation of galaxies, their structures, the different types of galaxies and the various dynamics that act upon them, including our very own Milky Way, are the topics that are discussed in-depth in this chapter.

A galaxy is a group of stars, clouds of gas, and dust particles that move together through the universe. There are billions of galaxies in the universe. They probably formed billions of years ago, soon after the universe began.

The Sun, Earth, and the other planets of the solar system are part of the Milky Way galaxy. Most galaxies move through the universe as part of clusters, or groups, of galaxies. The Milky Way is at one end of a cluster called the Local Group. It includes about 40 galaxies. The largest galaxy in the Local Group is the Andromeda Galaxy.

Even the smaller galaxies are made up of millions of stars. These galaxies may be 5,000 light-years across. (A light-year is the distance light travels in a year—about 5.8 trillion miles, or 9.5 trillion kilometers.) The Milky Way is roughly 20 times larger than that. It includes hundreds of billions of stars. The largest galaxies are even huger and contain trillions of stars.

Galaxies are divided into three main categories based on their shapes: spiral, elliptical, and irregular. Spiral galaxies look like pinwheels. They have a central disk and two or more curved arms winding outward. The disk consists mainly of older stars, while the arms have younger stars. The arms contain much dust and gas, from which new stars can form. The Milky Way is a spiral galaxy.

An elliptical galaxy may be shaped like a round ball or may be more stretched out, like an American football. It has more stars near its center and fewer stars in its outer regions. Elliptical galaxies vary widely in size. Small elliptical galaxies are the most common kind of galaxy in the universe.

Unusually shaped galaxies are called irregular galaxies. They are not very common. Many of them have no obvious center point. Some of them appear to be the result of a crash between several galaxies.

Types of Galaxies

There are three main types of galaxies:

- Elliptical
- Spiral
- Irregular

Elliptical Galaxies

Elliptical galaxies are among the largest single star systems in the universe and are built by the collisions of many smaller galaxies. Galaxies come in many shapes and sizes. Among the largest are the "elliptical", massive ball-shaped conglomerations of up to a trillion stars. Where they live and how they are built can teach astronomers a lot about the evolution of the universe and what the future has in store for our own Milky Way.

Elliptical are one of three main classes of galaxies defined by American astronomer Edwin Hubble in 1936. Unlike the other two classes—spirals and lenticulars—elliptical have almost no discernible structure. They are about as simple as a gathering of stars can be: massive blobs roughly spherical in shape. Through an amateur telescope, they appear as no more than a round, fuzzy patch of light against the dark background of the night sky.

The giant elliptical galaxy NGC 1132 sits about 300 million light-years from Earth. With a roughly 240,000 light-year diameter, it's over twice the size of our Milky Way! The many points of light surrounding it are a swarm of globular clusters which may be some of the remnants from billions of years of galactic collisions.

Peering deep into the core of the ellptical galaxy NGC 7052, astronomers have uncovered a disk of material thousands of light-years across. The material is spinning so fast that only a black hole with the mass of many millions of suns can be holding it all together.

One of the most intriguing aspects of giant elliptical is what they hide in their cores. Deep in the centers, elliptical each contains a supermassive black hole. A typical black hole forms from the death of a massive star and weighs at most a few times more than our sun. The black holes in galactic centers, however, can contain the mass of several billion suns. No single star can do that. The formation of these supermassive black holes is an area of active research. Most likely, they reflect the formation history of the galaxy. Each galactic collision funnels material down to the center of the galaxy where individual star-sized black holes can merge and grow.

Classification and Characteristics

In 1926, Edwin Hubble devised a system to classify galaxies. Known as the Hubble sequence, or the "Hubble tuning-fork," it organizes galaxies based on their shape. Under this organization, elliptical galaxies are classified by how stretched out they are. Galaxies classified as E0 appear to be almost perfect circles (remember, a circle is an ellipse), while those listed as E7 seem much longer than they are wide. A galaxy's appearance is related to how it lies on the sky when viewed from Earth. A galaxy having the E7 shape but seen head on would appear as an E0, for instance, because observers would not see its stretched shape, which lies "behind" it.

Elliptical galaxies have a broader range in size than other types of galaxies. The smallest are dwarf elliptical galaxies, which can be less than 10 percent of the size of the Milky Way. But elliptical can also stretch to more than a million light-years across, and contain more than ten trillion stars. M87, identified as one of the largest galaxies in the universe, is classified as an E0 elliptical galaxy.

Astronomers have identified more spiral galaxies than elliptical, but that's simply because the spirals are easier to spot. While spiral galaxies are bright, elliptical galaxies are dim. Spiral galaxies are hotbeds of star formation, but elliptical galaxies aren't nearly as prolific because they contain less gas and dust, which means fewer new (and brighter) stars are born. The existing stars inside an elliptical galaxy tend to be older, giving off more red light than younger stars.

So, why do astronomers think elliptical galaxies dominate the sky? Because when specific regions of the sky are studied in depth, more elliptical galaxies appear. Astronomers think such counts are consistent throughout the universe.

An elliptical galaxy, full of dark lanes of gas, likely formed in the merger of two other galaxies. Elliptical galaxies host less (or no) star birth than spiral galaxies like the Milky Way.

Because elliptical galaxies contain older stars and less gas, scientists think that they are nearing the end of the evolutionary line for galaxies. The universe is a violent place, and collisions between galaxies are frequent – indeed, the Milky Way is due to crash into the Andromeda Galaxy in a few billion years. When two spirals collide, they lose their familiar shape, morphing into the less-structured elliptical galaxies.

A supermassive black hole is thought to lie at the center of these ancient galaxies. These gluttonous giants consume gas and dust, and may play a role in the slower growth of elliptical galaxies.

Born from collision, elliptical galaxies are more commonly found around clusters and groups of galaxies. They are less frequently spotted in the early universe, which supports the idea that they evolved from the collisions that came later in the life of a galaxy.

Spiral Galaxies

Spirals are characterized by circular symmetry, a bright nucleus surrounded by a thin outer disk, and a superimposed spiral structure. They are divided into two parallel classes: normal spirals and barred spirals. The normal spirals have arms that emanate from the nucleus, while barred spirals have a bright linear feature called a bar that straddles the nucleus, with the arms unwinding from the ends of the bar. The nucleus of a spiral galaxy is a sharp-peaked area of smooth texture, which can be quite small or, in some cases, can make up the bulk of the galaxy. Both the arms and the disk of a spiral system are blue in colour, whereas its central areas are red like an elliptical galaxy. The normal spirals are designated S and the barred varieties SB. Each of these classes is sub classified into three types according to the size of the nucleus and the degree to which the spiral arms are coiled. The three types are denoted with the lowercase letters a, b, and c. There also exist galaxies that are intermediate between elliptical and spirals. Such systems have the disk shape characteristic of the latter but no spiral arms. These intermediate forms bear the designation So.

The Pinwheel Galaxy (M101), as seen in an optical image taken by the Hubble Space Telescope.

So Galaxies

These systems exhibit some of the properties of both the elliptical and the spirals and seem to be a bridge between these two more common galaxy types. Hubble introduced the So class long after his original classification scheme had been universally adopted, largely because he noticed the dearth of highly flattened objects that otherwise had the properties of elliptical galaxies. Sandage's elaboration of the So class yielded the characteristics described here.

So galaxy NGC 4753 in an optical image taken by the Sloan Digital Sky Survey Telescope at the Apache Point Observatory in Sunspot, N.M. NGC 4753 is a member of the Virgo Cluster of galaxies.

So galaxies have a bright nucleus that is surrounded by a smooth, featureless bulge and a faint outer envelope. They are thin; statistical studies of the ratio of the apparent axes (seen projected onto the sky) indicate that they have intrinsic ratios of minor to major axes in the range 0.1 to 0.3. Their structure does not generally follow the luminosity law of elliptical galaxies but has a form more like that for spiral galaxies. Some So systems have a hint of structure in the envelope, either faintly discernible arm like discontinuities or narrow absorption lanes produced by interstellar dust. Several So galaxies are otherwise peculiar, and it is difficult to classify them with certainty. They can be thought of as peculiar irregular galaxies (i.e., Irr II galaxies) or simply as some of the 1 or 2 percent of galaxies that do not fit easily into the Hubble scheme. Among these are such galaxies as NGC 4753, which has irregular dust lanes across its image, and NGC 128, which has a double, almost rectangular bulge around a central nucleus. Another type of peculiar So is found in NGC 2685. This nebula in the constellation Ursa Major has an apparently edge-on disk galaxy at its center, with surrounding hoops of gas, dust, and stars arranged in a plane that is at right angles to the apparent plane of the central object.

Sa Galaxies

These normal spirals have narrow, tightly wound arms, which usually are visible because of the presence of interstellar dust and, in many cases, bright stars. Most of them have a large amorphous bulge in the center, but there are some that violate this criterion, having a small nucleus around which is arranged an amorphous disk with superimposed faint arms. NGC 1302 is an example of the normal type of Sa galaxy, while NGC 4866 is representative of one with a small nucleus and arms consisting of thin dust lanes on a smooth disk.

The Sombrero Galaxy (M104), which is classified as an Sa/Sb galaxy, in an optical image taken by the Hubble Space Telescope.

Sb Galaxies

This intermediate type of spiral typically has a medium-sized nucleus. Its arms are more widely spread than those of the Sa variety and appear less smooth. They contain stars, star clouds, and interstellar gas and dust. Sb galaxies show wide dispersions in details in terms of their shape. Hubble and Sandage observed, for example, that in certain Sb galaxies the arms emerge at the nucleus, which is often quite small. Other members of this subclass have arms that begin tangent to a bright, nearly circular ring, while still others reveal a small, bright spiral pattern inset into the nuclear bulge. In any of these cases, the spiral arms may be set at different pitch angles. (A pitch angle is defined as the angle between an arm and a circle centered on the nucleus and intersecting the arm.)

Sb galaxy NGC 891 as seen edge-on in an optical image taken with the WIYN Telescope at the Kitt Peak National Observatory in Arizona, U.S. The dark band is made of interstellar dust.

Hubble and Sandage noted further deviations from the standard shape established for Sb galaxies. A few systems exhibit a chaotic dust pattern superimposed upon the tightly wound spiral arms.

Some have smooth, thick arms of low surface brightness, frequently bounded on their inner edges with dust lanes. Finally, there are those with a large, smooth nuclear bulge from which the arms emanate, flowing outward tangent to the bulge and forming short arm segments. This is the most familiar type of Sb galaxy and is best exemplified by the giant Andromeda Galaxy.

Many of these variations in shape remain unexplained. Theoretical models of spiral galaxies based on a number of different premises can reproduce the basic Sb galaxy shape, but many of the deviations noted above are somewhat mysterious in origin and must await more detailed and realistic modeling of galactic dynamics.

Sc Galaxies

These galaxies characteristically have a very small nucleus and multiple spiral arms that are open, with relatively large pitch angles. The arms, moreover, are lumpy, containing as they do numerous irregularly distributed star clouds, stellar associations, star clusters, and gas clouds known as emission nebulae.

As in the case of Sb galaxies, there are several recognizable subtypes among the Sc systems. Sandage has cited six subdivisions: galaxies, such as the Whirlpool Galaxy (M51), that have thin branched arms that wind outward from a tiny nucleus, usually extending out about 180° before branching into multiple segments, systems with multiple arms that start tangent to a bright ring centred on the nucleus, those with arms that are poorly defined and that span the entire image of the galaxy, those with a spiral pattern that cannot easily be traced and that are multiple and punctuated with chaotic dust lanes, those with thick, loose arms that are not well defined—e.g., the nearby galaxy M33 (the Triangulum Nebula)—and transition types, which are almost so lacking in order that they could be considered irregular galaxies.

Some classification schemes, such as that of the French-born American astronomer Gerard de Vaucouleurs, give the last of the above-cited subtypes a class of its own, type Sd. It also has been found that some of the variations noted here for Sc galaxies are related to total luminosity. Galaxies of the fifth subtype, in particular, tend to be intrinsically faint, while those of the first subtype are among the most luminous spirals known. This correlation is part of the justification for the luminosity classification discussed below.

SB Galaxies

The luminosities, dimensions, spectra, and distributions of the barred spirals tend to be indistinguishable from those of normal spirals. The subclasses of SB systems exist in parallel sequence to those of the latter.

There are SB0 galaxies that feature a large nuclear bulge surrounded by a dislike envelope across which runs a luminous featureless bar. Some SB0 systems have short bars, while others have bars that extend across the entire visible image. Occasionally there is a ring like feature external to the bar. SBa galaxies have bright, fairly large nuclear bulges and tightly wound, smooth spiral arms that emerge from the ends of the bar or from a circular ring external to the bar. SBb systems have a smooth bar as well as relatively smooth and continuous arms. In some galaxies of this type, the arms start at or near the ends of the bar, with conspicuous dust lanes along the inside of the bar that can be traced right up to the nucleus. Others have arms that start tangent to a ring external to the bar. In SBc galaxies, both the arms and the bar are highly resolved into star clouds and stellar associations. The arms are open in form and can start either at the ends of the bar or tangent to a ring.

Barred spiral galaxy NGC 1300.

Irregular Galaxies

An irregular galaxy is the catchall name given to any galaxy that does not neatly fit into one of the categories of the Hubble classification scheme. They have no defined shape or structure and may have formed from collisions, close encounters with other galaxies or violent internal activity. They contain both old and young stars, significant amounts of gas and usually exhibit bright knots of star formation.

Due to the diversity of objects that fall into this category it is difficult to constrain sizes, masses and luminosities. Dwarf irregulars can be as small as 3 kiloparsecs and contain as little as 108 solar masses of material. At the other end of the scale, the larger irregulars can be up to 10 kiloparsecs across and contain 1010 solar masses of material. Their luminosities range from 107 to 109 solar, making them generally fainter than spiral galaxies.

The best known examples of irregular galaxies are the Small and Large Magellanic clouds. These are companion galaxies to our own Milky Way, and can be easily seen at dark sites in the Southern Hemisphere.

Galactic Formation

Galaxy formation is a core theme of cosmology. Galaxies are the basic unit of cosmology. They contain stars, gas, dust and alot of dark matter. They are the only `signposts' from here to the edge of the Universe and contain the fossil clues to this earlier time.

The physics of galaxy formation is complicated because it deals with the dynamics of stars (gravitational interaction), thermodynamics of gas and energy production of stars. For example, stars form from gas clouds, but new stars heat these clouds, which dissipates them and stops the formation of other stars.

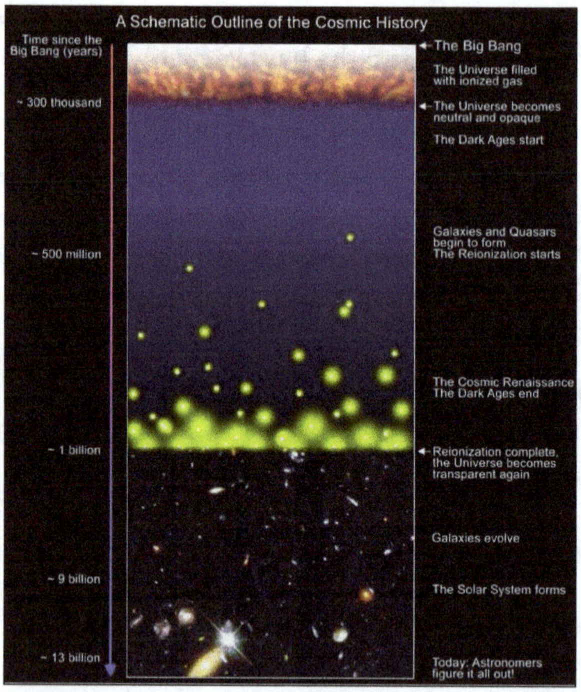

Protogalaxies

After recombination, density enhancements either grew or dispersed. According to our hybrid top-down/bottom-up scenario, an assortment of enhancements formed of various sizes. Small, dense ones collapsed first, large ones formed slower and fragmented as they collapsed.

The first lumps that broke free of the Universe's expansion were mostly dark matter and some neutral hydrogen with a dash of helium. Once this object begins to collapse under its own gravity, it is called a protogalaxy. The first protogalaxies appeared about 14 billion years ago.

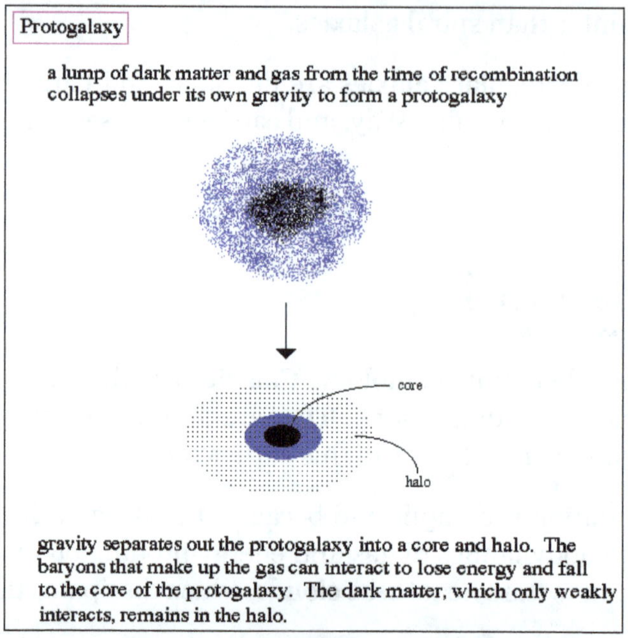

Note that dark matter and ordinary matter (in the form of hydrogen and helium gas at this time) separate at this time. Gas can dissipate its energy through collisions. The atoms in the gas collide and heat up, the heat is radiated in the infrared (light) and the result is the gas loses energy, moves slowly = collapses to the center. Dark matter does not interact this way and continues to orbit in the halo.

Even though there are no stars yet, protogalaxies should be detectable by their infrared emission (i.e. their heat). However, they are very faint and very far away (long time ago), so our technology has not been successful in discovering any at this time.

Formation of the First Stars

As the gas in the protogalaxy loses energy, its density goes up. Gas clouds form and move around in the protogalaxy on or bits. When two clouds collide, the gas is compressed into a shock front.

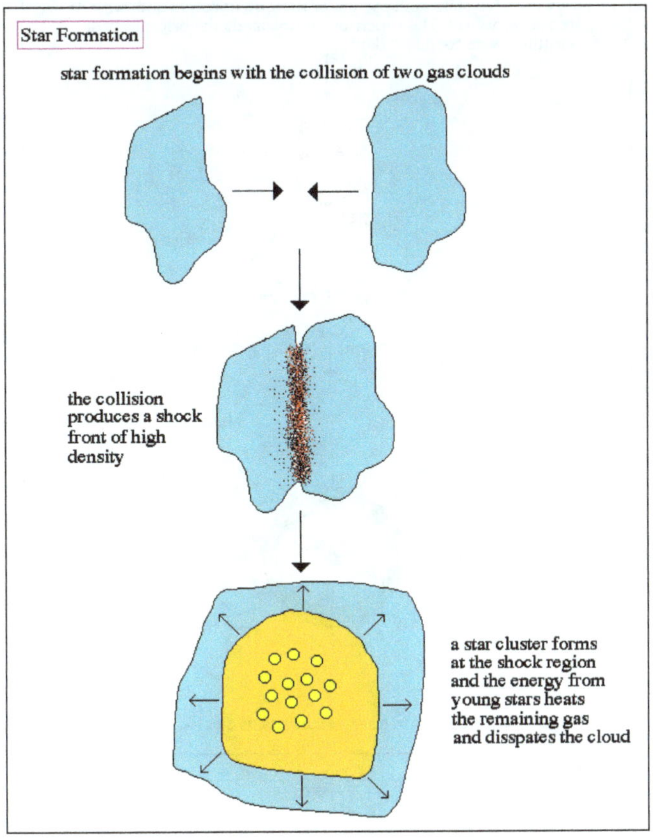

The first stars in a galaxy form in this manner. With the production of its first photons by thermonuclear fusion, the galaxy becomes a primeval galaxy. Star formation sites in primeval galaxies are similar to star forming regions in present-day galaxies. A grouping of young stars embedded in a cloud of heated gas. The gas will eventually be pushed away from the stars to leave a star cluster.

The first stars in our Galaxy are the globular star clusters orbiting outside the stellar disk which contains the spiral arms. Most galaxies with current star formation have an underlying distribution of old stars from the first epoch of star formation 14 billion years ago.

Stellar Death

The most massive stars end their lives as supernova, the explosive destruction of a star. Supernovas occur when a star uses up its interior fuel of hydrogen and collapses under its own weight. The infalling hydrogen from the star's outer envelope hits the core and ignites explosively.

During the explosion, runaway fusion occurs and all the elements in the periodic table past lithium are produced. This is the only method of producing the heavy elements and is the origin to all the elements in your body. This shell of enriched gas is ejected into the galaxy's gas supply. Thus, the older a galaxy, the richer its gas is in heavy elements, a process called chemical evolution.

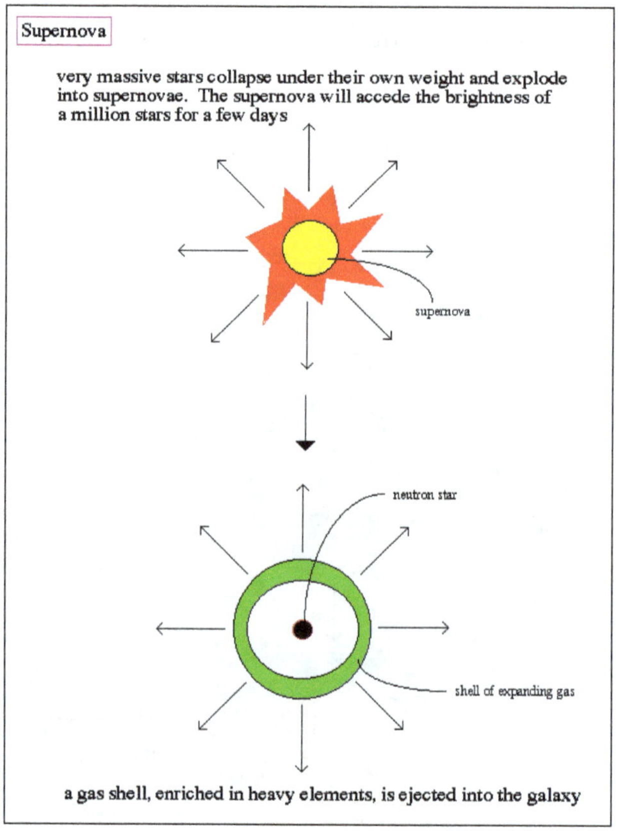

Ellipticals vs. Spirals

The two most distinct galaxy types are ellipticals and spirals. Ellipticals have no ongoing star formation today, spirals have a lot. Assuming that ellipticals and spirals are made from the same density enhancements at the time of recombination, why did they evolve into very difference appearances and star formation rates?

The answer is how rapid their initial star formation was when they formed. If star formation proceeds slowly, the gas undergoes collisions and conservation of angular momentum forms a disk (a spiral). If star formation is rapid and all the gas is used up in an initial burst, the galaxy forms as a smooth round shape an elliptical.

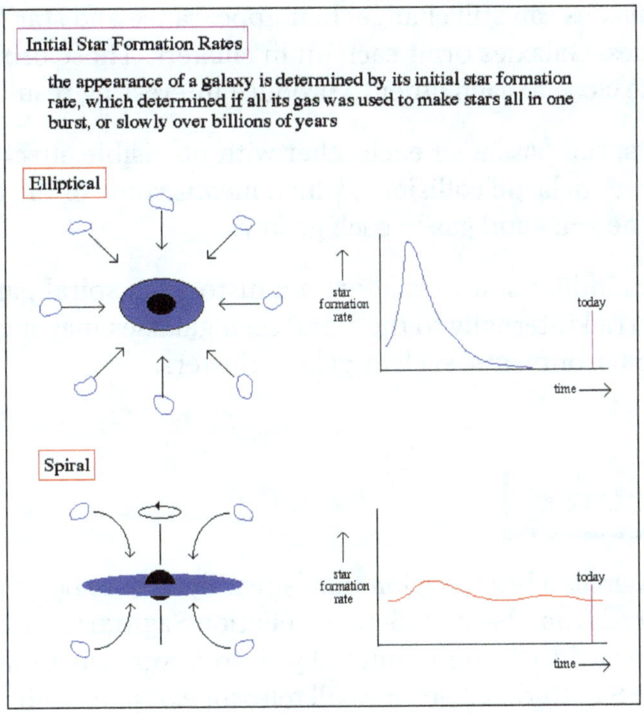

Gas falling into a spiral disk is slowed by collisions and star formation continues till today. The spiral arms and patterns are due to ongoing star formation, whereas ellipticals used all their gas supplies in an initial burst 14 billion years ago and now have no ongoing star formation.

Galaxy Mergers/Interactions

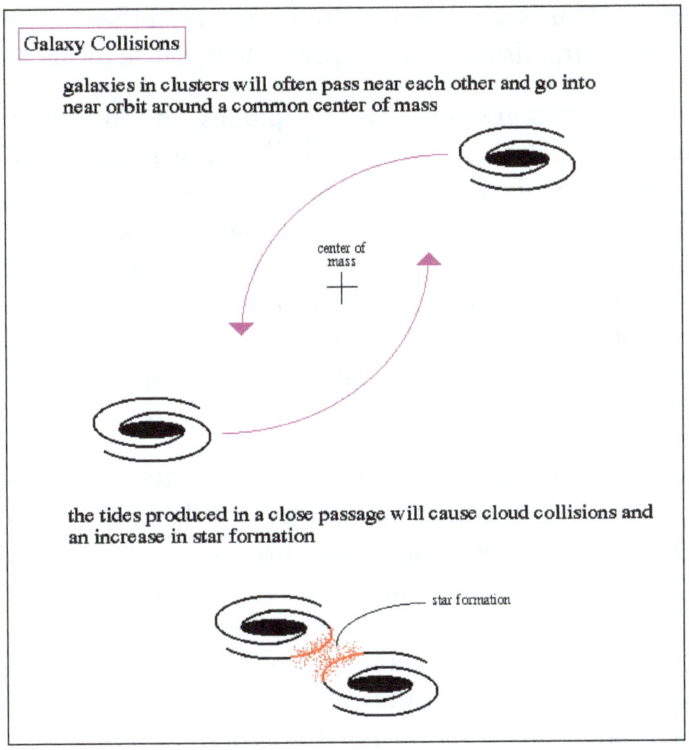

After their formation, galaxies can still change their appearance and star formation rates by interactions with other galaxies. Galaxies orbit each on in clusters. Those orbits can sometimes cause two galaxies to pass quite close to each other to produce interesting results.

Solid objects, like planets, can pass near each other with no visible effects. However, galaxies are not solid, and can undergo inelastic collisions, which means some of the energy of the collision is transfered internally to the stars and gas in each galaxy.

The tidal forces will often induce star formation and distort the spiral pattern in both galaxies. If enough energy is transferred internally to the stars, then galaxies may merge. Galaxy mergers are most frequent in dense environments, such as galaxy clusters.

Galactic Structure

Passing around the sky there is a broad region that is readily seen to be brighter than the rest of the night sky. It has been traced from the summer constellation Sagittarius northward through Cyngus into Perseus, then southward to Orion (winter sky) into Centaurus (Southern Hemisphere sky) then back northward into Sagittarius. Even a small telescope or pair of binoculars reveals this band to be bright because of the cumulative effect of millions of faint stars. This is the Milky Way. That it is due to myriads of faint stars distributed in a great circle about the position of the Sun shows the Galaxy's basic structure, the manner in which the stars and interstellar material that make up the Galaxy are distributed in space, is flat. This is the plane of the Galaxy, where the greater part of the stars and interstellar material exist. The brightest part of Milky Way, visible low on the southern horizon in the summer sky toward the constellation of Sagittarius, is bright because the star density increases in this direction. This is the direction to the center of the Galaxy, though starlight coming from the vast bulk of stars in this direction is invisible because of the absorption by the dust.

The distribution of dusty, absorption nebulae is very patchy, and there are "windows," directions passing close to the center in which there is relatively little absorption, that allow study of the distant stars. In these directions and elsewhere in the halo of the Galaxy, the distribution of RR Lyrae and other stars yields its density structure. In the same manner, the directions and distances to the globular clusters may be mapped in three dimensions. The clusters are concentrated in the direction of Sagittarius, and their density decreases outward, allowing astronomers to outline the outer structure of the Galaxy. From their distribution, the position of the densest part of the Galaxy, the center, may be determined. The galactocentric distance of the Sun is currently estimated as $R_{\odot} \approx$ 8 Kpc (25,000 ly).

The brightest stars at the center of the Galaxy may also be studied using long wavelength infrared radiation. The total extent of the plane of the Galaxy can be deduced by analyzing observations of the 21-centimeter radiation of neutral hydrogen 360° around the plane. This analysis gives the size of the whole Galaxy as about 30,000 pc diameter (100,000 ly). Scans in 21-cm above and below the plane, together with observations of stars perpendicular to the plane, give a total thickness of about 500 pc (1,600 ly), with half the gas mass within 110 pc (360 ly) of the center of the plane. Radio studies also reveal that the fundamental plane of the Galaxy is warped, like a fedora hat, with the brim pushed up on one side and down on the other.

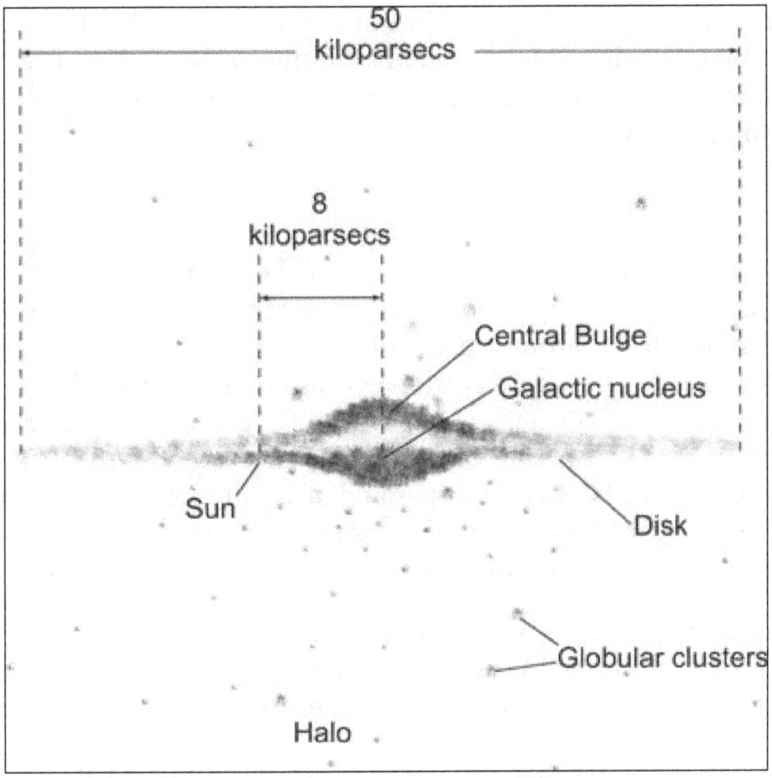

An external view of the Milky Way, looking edge-on or sideways into the disk.

It is bent down on Sun side of the Galaxy and up on the opposite side, due to a gravitational resonance with the Magellanic Clouds, which move in an orbit about the Milky Way.

While the greater part of the mass of the Milky Way lies in the relatively thin, circularly symmetric plane or disk, there are three other recognized components of the Galaxy, each marked by distinct patterns of spatial distribution, motions, and stellar types. These are the halo, nucleus, and corona.

Disk

The disk consists of those stars distributed in the thin, rotating, circularly symmetric plane that has an approximate diameter of 30,000 pc (100,000 ly) and a thickness of about 400 to 500 pc (1,300 to 1,600 ly). Most disk stars are relatively old, although the disk is also the site of present star formation as evidenced by the young open clusters and associations. The estimated present conversion rate of interstellar material to new stars is only about 1 solar mass per year. The Sun is a disk star about 8 kpc (25,000 ly) from center. All these stars, old to young, are fairly homogeneous in their chemical composition, which is similar to that of the Sun.

The disk also contains essentially the entire Galaxy's content of interstellar material, but the gas and dust are concentrated to a much thinner thickness than the stars; half the interstellar material is within about 25 pc (80 ly) of the central plane. Within the interstellar material, denser regions contract to form new stars. In the local region of the disk, the position of young O and B stars, young open clusters, young Cepheid variables, and HII regions associated with recent star formation reveal that star formation does not occur randomly in the plane but in a spiral pattern analogous to the spiral arms found in other disk galaxies.

The disk of the Galaxy is in dynamical equilibrium, with the inward pull of gravity balanced by motion in circular orbits. The disk is fairly rapidly rotating with a uniform velocity about 220 km. Over most of the radial extent of the disk, this circular velocity is reasonably independent of the distance outward from the center of the Galaxy.

Halo and Bulge

Some stars and star clusters (globular clusters) form the halo component of the Galaxy. They surround and interpenetrate the disk, and are thinly distributed in a more or less spherical (or spheroidal) shape symmetrically around the center of the Milky Way. The halo is traced out to about 100,000 pc (325,000 ly), but there is no sharp edge to the Galaxy; the density of stars simply fades away until they are no longer detectable. The halo's greatest concentration is at its center, where the cumulative light of its stars becomes comparable to that of the disk stars. This region is called the (nuclear) bulgeof the Galaxy; its spatial distribution is somewhat more flattened than the whole halo. There is also evidence that the stars in the bulge have slightly greater abundances of heavy elements than stars at greater distances from the center of the Galaxy.

The halo stars consist of old, faint, red main sequence stars or old, red giant stars, considered to be among the first stars to have formed in the Galaxy. Their distribution in space and their extremely elongated orbits around the center of the Galaxy suggest that they were formed during one of the Galaxy's initial collapse phases. Forming before there had been significant thermonuclear processing of materials in the cores of stars, these stars came from interstellar matter with few heavy elements. As a result, they are metal poor. At the time of their formation, conditions also supported the formation of star clusters that had about 10^6 solar masses of material, the globular clusters. Today there exists no interstellar medium of any consequence in the halo and hence no current star formation there. The lack of dust in the halo means that this part of the Galaxy is transparent, making observation of the rest of the universe possible.

Halo stars can easily be discovered by proper motion studies. In extreme cases, these stars have motions nearly radial to the center of the Galaxy—hence at right angles to the circular motion of the Sun. Their net relative motion to the Sun therefore is large, and they are discovered as high-velocity stars, although their true space velocities are not necessarily great. Detailed study of the motions of distant halo stars and the globular clusters shows that the net rotation of the halo is small. Random motions of the halo stars prevent the halo from collapsing under the effect of the gravity of the whole Galaxy.

Nucleus

The nucleus is considered to be a distinct component of the Galaxy. It is not only the central region of the Galaxy where the densest distribution of stars (about 50,000 stars per cubic parsec compared to about 1 star per cubic parsec in the vicinity of the Sun) of both the halo and disk occurs, but it is also the site of violent and energetic activity. The very center of the Galaxy harbors objects or phenomena that are not found elsewhere in the Galaxy. This is evidenced by a high flux of infrared, radio, and extremely short wavelength gamma radiation coming from the center, a specific infrared source known as Sagittarius A. Infrared emissions in this region show that a high density of cooler stars exists there, in excess of what would be expected from extrapolating the normal distribution of halo and disk stars to the center.

The nucleus is also exceptionally bright in radio radiation produced by the interaction of high-velocity charged particles with a weak magnetic field (synchrotron radiation). Of greater significance is the variable emission of gamma rays, particularly at energy of 0.5 MeV. This gamma-ray emission line has only one source—the mutual annihilation of electrons with anti-electrons, or positrons, the source of which in the center has yet to be identified. Theoretical attempts to explain these phenomena suggest a total mass involved of 10^6–10^7 solar masses in a region perhaps a few parsecs in diameter. This could be in the form of a single object, a massive black hole; similar massive objects appear to exist in the centers of other galaxies that show energetic nuclei. By the standards of such active galaxies, however, the nucleus of the Milky Way is a quiet place, although interpretations of the observed radiation suggest the existence of huge clouds of warm dust, rings of molecular gas, and other complex features.

Exterior to the Halo

The gravitational influence of the Galaxy extends to an even greater distance of about 500,000 pc (1,650,000 ly) (the late astronomer Bart Bok suggested this region could be called the corona of the Galaxy). In this volume there appears to be an excess of dwarf galaxies associated with the Milky Way, drawn into its proximity by its large gravitational pull. This includes the Magellanic Clouds, which lie in the debris of the Magellanic Stream. The Magellanic Stream consists of a band of hydrogen gas and other materials that extends around the Galaxy, marking the orbital path of these companion galaxies. The tidal gravitational field of the Galaxy apparently is ripping them apart, a process that will be completed in the next two to three billion years. This galactic cannibalism, the destruction of small galaxies, and the accretion of their stars and gas into a larger galactic object likely has happened in the past, perhaps many times. A second, small companion galaxy in the direction of Sagittarius (the Sagittarius galaxy) appears to be another victim of this process. Like the Magellanic Clouds, its stars and interstellar material will ultimately be incorporated into the body of the Milky Way. The total number of dwarf galaxies near the Milky Way is about a dozen and includes objects such as Leo I, Leo II, and Ursa Major. A similar cloud of dwarf galaxies exists about the Andromeda Galaxy.

Rotation Curve of the Galaxy

An alternative means of studying the structure of the Galaxy, complementary to looking at the distribution of specific objects, is to deduce the total distribution of mass. This may be done by analyzing the rotation curve, or the circular velocity V(R) of the disk objects moving around the center of the Galaxy as a function of the distance R out from the center. A check on the accuracy of the deduced motion in the Galaxy is given by the rotation curves of similar galaxies, which would be expected to rotate in the same basic fashion. Like the Milky Way, the rotations of other galaxies show a linear increase of velocity near their centers rising to a maximum value and then becoming basically constant over the remainder of the disk.

Determination of V(R) from within the Galaxy is not as straightforward as measuring the rotation of another galaxy that is observed from outside. Observation of neighboring stars or of interstellar gas gives only relative motions. Thus, calculating the absolute solar velocity involves first looking at nearby galaxies and determining what direction the Sun appears to be moving in.

The Sun and its neighboring stars are found to be moving about the center of the Galaxy with a speed of 220 km/s in the direction of the northern constellation Cygnus, at a right angle to the direction towards the center. In the galactic coordinate system used by astronomers, this movement is toward a galactic longitude of 90°. Sweeping around the Galaxy in its plane, galactic longitude starts at 0° toward the center, increases to 90° in the direction of rotation (Cygnus), to 180° in the anti-center direction (Orion), to 270° in the direction from which the Sun moves (Centaurus), and finally to 360° when the direction of the center is again reached. Use of Doppler shifts and proper motions applied to stars near the sun provide some idea of the local rotation curve; nearby disk stars on average appear to move in circular orbits about the center with the same circular velocity as the Sun. The interstellar dust prevents study by optical techniques of the rest of the Galaxy; thus, the 21-centimeter radiation of neutral hydrogen must be used to determine its pattern of motion. Again, the Doppler Shift gives only a relative or line-of-sight velocity for the gas anywhere in the Galaxy, but knowledge of the solar velocity and geometry allows calculation of the velocity at other radii from the galactic center.

The rotation curve of the Galaxy shows that it does not rotate as a solid disk (velocity directly proportional to distance out from the rotational axis). Rather, the rotational velocity is more or less constant over most of the disk.

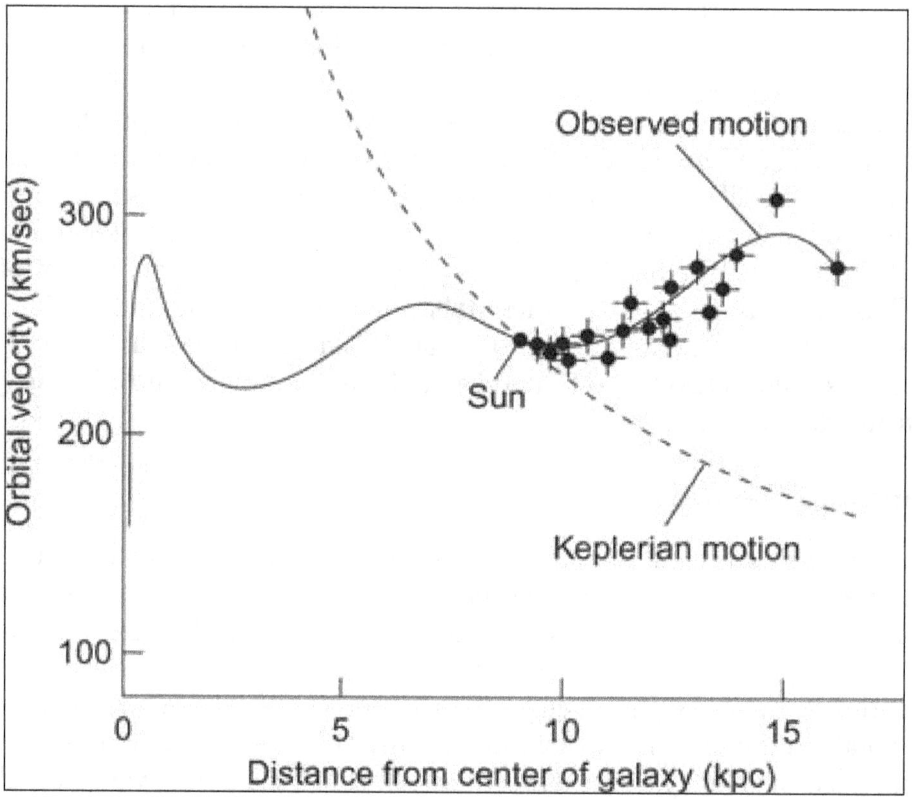

Rotation curve of the Galaxy. If the greatest portion of the mass of the Galaxy were concentrated at its center, then orbital motions would rapidly decrease with radius (dashed line) in the manner of the planetary motions about the Sun described by Kepler.

Viewed as a giant race course, this means that on average all-stars move the same distance in a given amount of time, but because the circular paths of outer stars are larger than those closer to the center, the outer stars slip progressively behind the inner stars. This effect is called differential rotation, and it has significant effects on the distribution of star-forming regions; any large star-forming region will be sheared into a spiral arc. If the Galaxy rotated as a solid disk, there would be no differential rotation.

Stars, including the Sun, have small components of motion that deviate from a pure circular motion about the center of the Galaxy. This peculiar motion for the Sun is about 20 km/s, a small drift in the general direction of the bright summer star Vega. This results in an approximate 600 pc (1900 ly) in-and-out deviation from a true circular orbit as the Sun orbits the center of the Galaxy with a period of 225 million years. A second consequence is an oscillation, with a much shorter period of about 60 million years, up and down through the plane of the disk. In other words, the Sun moves up and down about four times during each trip around the center of the Galaxy. This oscillation has amplitude of 75 pc (250 ly). At present, the Sun is 4 pc (13 ly) above the galactic plane, moving upward into the Galaxy's Northern Hemisphere.

Mass Distribution

In one sense, the Galaxy is analogous to the solar system: The flatness is the result of the operation of the same physical laws. As the material of both contracted at their time of formation, conservation of angular momentum resulted in increased rotational velocities until a balance against gravity was achieved in an equatorial plane. Material above or below that plane continued to fall inwards until the mass distribution became flat. In specific detail, the mass distributions are very dissimilar. The mass of the Galaxy is distributed through a large volume of space, whereas the mass of the solar system is essentially only that of the Sun and is located at the center. The flat disk of the Galaxy implies that rotation plays the dominant role in the balance against gravitation, which, in turn, depends on the mass distribution. The mass M(R) as a function of radius R is determined by applying a modification of Kepler's Third Law to the rotation curve V(R), to obtain,

$$M(R) \approx RV^2(R)/G$$

where G is the gravitational constant. Thus, astronomers can determine the mass structure of the Galaxy. Its total mass may be as great as 10^{12} solar masses.

Because the mass in the Galaxy is distributed over a large volume, the pattern of rotation differs from that in the solar system. For the planets, orbital velocities decrease with radial distance outward, $V(R) \propto R^{-1/2}$ (Keplerian motion); in the Galaxy, the circular velocity rises linearly $V(R) \propto R$ near the center, and then is relatively unchanging over the remainder of the disk, $V(R) \propto$ constant. This form of rotation curve implies a relatively constant mass density near the center; but further out, the density decreases inversely with the square of the radius.

The motions of the stars are also affected by the spatial distribution of the mass. The nature of Newtonian gravity is that a circularly or spherically symmetric mass distribution always exerts a force toward the center, but this force depends only on that part of the mass that is closer to the center than the object that feels the force. If a star moves outward in the Galaxy, it feels the gravitational force from a larger fraction of the total mass; when it moves closer to the center, less of the mass is exerting a force on the object. As a result, orbits of stars are not closed ellipses like those of the planets, but instead more closely resemble the patterns produced by a Spiro graph. Additionally, a planetary orbit is a flat plane; hence, if that orbit is inclined to the overall plane of the solar system, in one complete circuit about the Sun the planet moves once above and once below the solar system plane. A star, however, will oscillate up and down several times in one passage around the center of the Galaxy.

Spiral Arm Phenomenon

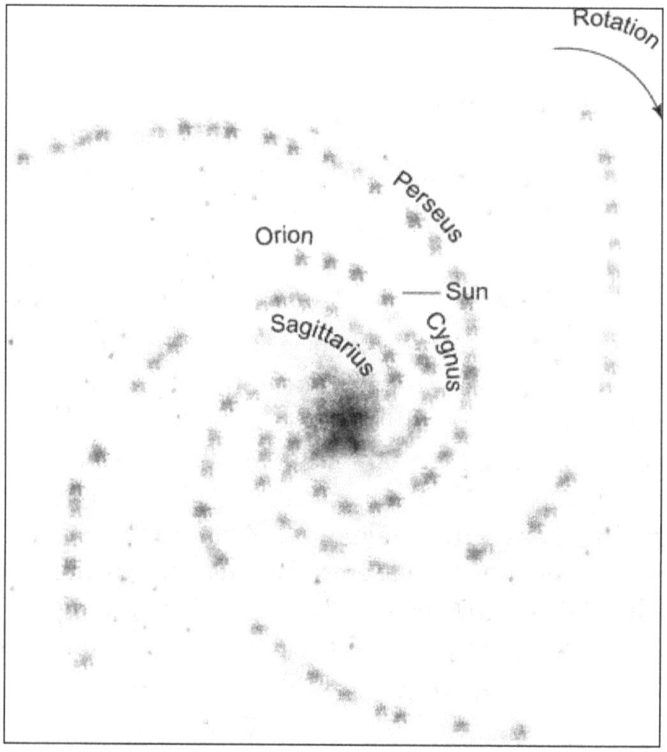

A schematic interpretation of the spiral features in the disk of the Milky Way Galaxy. The various spiral arms are named after the constellations in which directions their brightest features are observed.

In the Galaxy, the mass structure of the disk is not perfectly smooth. Instead, there are regions in the disk where the density of stars is slightly larger than the average. In these same regions, the density of the interstellar material may be significantly larger. These density variations, or fluctuations, are not completely random; they show a global pattern of spirality, or spiral arms, within the disk. Again the dust in our Galaxy is a problem; thus, spiral features easily studied in distant disk galaxies can give us insight to the pattern in the Milky Way. Stellar and no stellar objects associated with the spiral arms can be mapped out only locally in our Galaxy, out to 3 kpc (10,000 ly) or so, because in regions of higher density of interstellar material, star formation occurs. In particular, the brightest O and B stars are indicative of the most recent star formation. They and other objects associated with recent star formation (emission regions, Cepheid variables, young star clusters) may be used as optical tracers of the spiral arm pattern. Analysis of 21-centimeter observations is more difficult, but suggests that coincident with young stellar objects are the denser regions of interstellar material.

To have a pattern of compression (higher density) and rarefaction (lower density) in the spiral arm pattern that exists over the whole disk of a galaxy requires energy, in the same manner that the sound produced when a person speaks requires energy. Both phenomena are examples of wave phenomena. A sound wave is a pattern of alternate compression and rarefaction in air molecules.

Like any wave phenomena, the energy that is responsible for the wave will dissipate into random motions, and the wave pattern should die away in a relatively short period of time.

The density wave that passes through the disk of the Galaxy can be better related to the density waves that are found on freeways. At times, any given driver will be in the midst of "traffic," but at other times, he or she will seem to be the only driver on the road. Physically, these waves are the result of two factors. First, not all automobiles are driven at the same speed. There are slower and faster drivers. Second, congestion occurs because there are a limited number of lanes for the traffic flow. Faster drivers come up from behind and are delayed as they weave from lane to lane in their effort to get through to the head of the pack and resume their higher speed. They then can rush ahead, only to get caught up in the next pattern of congestion. Slower drivers get left behind until the next traffic wave catches up to them. Seen from a helicopter, a wave of alternatively denser and thinner distributions of cars is traveling down the highway; those cars in the dense regions, however, change as the faster cars move through and the slower ones drift behind.

In the Galaxy, the dynamics are slightly different in that the "highway" is a circulation about a galactic center, and the congestion is due to the stronger gravity in regions with larger numbers of stars. The spiral density wave theory begins by postulating the existence of a spirally structured pattern of density enhancement in a galactic disk. In the regions of extra density, the extra gravity affects motions and causes the gas and stars to "pile up" momentarily in these spirally shaped regions. Once the stars have passed through the spiral arm, they can move slightly faster until they catch up to the next spiral arm where they again will be momentarily delayed. The gas particles, being much less massive than the stars, are significantly more affected by the excess gravity and can be compressed to five times the average density of the interstellar matter in the disk. This compression is enough to trigger star formation; the newly formed luminosity O and B stars and their associated emission regions thus light up the regions of the spiral arms. The theory very successfully shows that a spiral density enhancement in the form of two well-formed spiral arms, a so-called Grand Design, is self-sustaining for several rotations of a galaxy. In the Milky Way, the expected flow pattern in stellar motions due to acceleration by the gravity of the spiral arms, superimposed on the overall circular motion about the center of the Galaxy, has been observed.

The evidence for the excitation of the wave in the first place should be evident because the lifetime of such a wave is rather short (a few galaxy rotation periods). In fact, a Grand Design spiral galaxy is generally accompanied by a companion galaxy whose recent close passage by the larger galaxy gave the gravitational stimulus to produce the density wave.

Not all galaxies show a distinct, two-armed spiral pattern. In fact, the majority of disk galaxies show numerous arc-like features, apparent fragments of spiral features that are referred to as flocculent galaxies. Each arc represents regions lit up by the bright stars of recent star formation and are explained by the stochastic self-propagating star formation theory. Given an initial collapse of interstellar gas into a group of stars, in due course a massive star will undergo a supernova explosion. Shock waves moving outward then push the ambient interstellar material into denser condensations and can trigger a next generation of new stars. If there are new massive stars, there will be subsequent supernovae, and the process repeats (the self-propagating aspect). This cycle continues until the interstellar gas is depleted, or until by chance no new massive stars form (this is the random, or stochastic, aspect of this theory). During the existence of a wave of star formation moving outwards from some original position, however, the growing region of star formation is affected by differential rotation in the disk; the outer part of the star-forming region lags behind the inner part. The region of star formation is therefore smeared into a spiral arc, as would be all other growing, star-forming regions elsewhere in the disk; but there would be no grand design.

5

Stars and Stellar Astrophysics

Stars are the celestial bodies made out of hydrogen and helium that produce light and heat from the continuous nuclear reaction taking place at their cores. Star formation and stellar evolution are topics actively pursued by scientist in order to understand these astronomical objects better, given that our sun is also a star. This chapter covers all matters related to stars and stellar astrophysics comprehensively.

Star

A star-forming region in the Large Magellanic Cloud.

False-color imagery of the Sun, a G-type main-sequence star, the closest to Earth.

A star is a luminous sphere of plasma held together by its own gravity. The nearest star to Earth is the Sun. Many other stars are visible to the naked eye from Earth during the night, appearing as a multitude of fixed luminous points in the sky due to their immense distance from Earth. Historically, the most prominent stars were grouped into constellations and asterisms, the brightest of which gained proper names. Astronomers have assembled star catalogues that identify the known stars and provide standardized stellar designations. However, most of the stars in the Universe, including all stars outside our galaxy, the Milky Way, are invisible to the naked eye from Earth. Indeed, most are invisible from Earth even through the most powerful telescopes.

For at least a portion of its life, a star shines due to thermonuclear fusion of hydrogen into helium in its core, releasing energy that traverses the star's interior and then radiates into outer space. Almost all naturally occurring elements heavier than helium are created by stellar nucleosynthesis during the star's lifetime, and for some stars by supernova nucleosynthesis when it explodes. Near the end of its life, a star can also contain degenerate matter. Astronomers can determine the mass, age, metallicity (chemical composition), and many other properties of a star by observing its motion through space, its luminosity, and spectrum respectively. The total mass of a star is the main factor that determines its evolution and eventual fate. Other characteristics of a star, including diameter and temperature, change over its life, while the star's environment affects its rotation and movement. A plot of the temperature of many stars against their luminosities produces a plot known as a Hertzsprung–Russell diagram (H–R diagram). Plotting a particular star on that diagram allows the age and evolutionary state of that star to be determined.

A star's life begins with the gravitational collapse of a gaseous nebula of material composed primarily of hydrogen, along with helium and trace amounts of heavier elements. When the stellar core is sufficiently dense, hydrogen becomes steadily converted into helium through nuclear fusion, releasing energy in the process. The remainder of the star's interior carries energy away from the core through a combination of radiative and convective heat transfer processes. The star's internal pressure prevents it from collapsing further under its own gravity. A star with mass greater than 0.4 times the Sun's will expand to become a red giant when the hydrogen fuel in its core is exhausted. it In some cases, it will fuse heavier elements at the core or in shells around the core. As the star expands it throws a part of its mass, enriched with those heavier elements, into the interstellar environment, to be recycled later as new stars. Meanwhile, the core becomes a stellar remnant: a white dwarf, a neutron star, or if it is sufficiently massive a black hole.

Binary and multi-star systems consist of two or more stars that are gravitationally bound and generally move around each other in stable orbits. When two such stars have a relatively close orbit, their gravitational interaction can have a significant impact on their evolution. Stars can form part of a much larger gravitationally bound structure, such as a star cluster or a galaxy.

Observation History

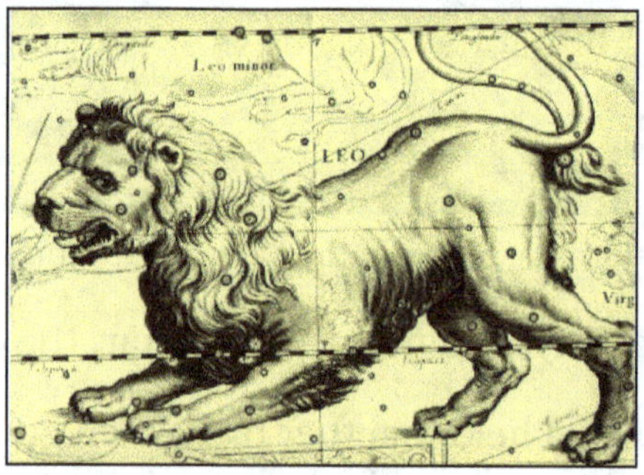

People have seen patterns in the stars since ancient times. This 1690 depiction of the constellation of Leo, the lion, is by Johannes Hevelius.

Historically, stars have been important to civilizations throughout the world. They have been part of religious practices and used for celestial navigation and orientation. Many ancient astronomers believed that stars were permanently affixed to a heavenly sphere and that they were immutable. By convention, astronomers grouped stars into constellations and used them to track the motions of the planets and the inferred position of the Sun. The motion of the Sun against the background stars (and the horizon) was used to create calendars, which could be used to regulate agricultural practices. The Gregorian calendar, currently used nearly everywhere in the world, is a solar calendar based on the angle of the Earth's rotational axis relative to its local star, the Sun.

The oldest accurately dated star chart was the result of ancient Egyptian astronomy in 1534 BC. The earliest known star catalogues were compiled by the ancient Babylonian astronomers of Mesopotamia in the late 2nd millennium BC, during the Kassite Period (*ca.* 1531–1155 BC).

The constellation of Leo as it can be seen by the naked eye. Lines have been added.

The first star catalogue in Greek astronomy was created by Aristillus in approximately 300 BC, with the help of Timocharis. The star catalog of Hipparchus (2nd century BC) included 1020 stars, and was used to assemble Ptolemy's star catalogue. Hipparchus is known for the discovery of the first recorded *nova* (new star). Many of the constellations and star names in use today derive from Greek astronomy.

In spite of the apparent immutability of the heavens, Chinese astronomers were aware that new stars could appear. In 185 AD, they were the first to observe and write about a supernova, now known as the SN 185. The brightest stellar event in recorded history was the SN 1006 supernova, which was observed in 1006 and written about by the Egyptian astronomer Ali ibn Ridwan and several Chinese astronomers. The SN 1054 supernova, which gave birth to the Crab Nebula, was also observed by Chinese and Islamic astronomers.

Medieval Islamic astronomers gave Arabic names to many stars that are still used today and they invented numerous astronomical instruments that could compute the positions of the stars. They built the first large observatory research institutes, mainly for the purpose of producing *Zij* star catalogues. Among these, the *Book of Fixed Stars* (964) was written by the Persian astronomer Abd al-Rahman al-Sufi, who observed a number of stars, star clusters (including the Omicron Velorum and Brocchi's Clusters) and galaxies (including the Andromeda Galaxy). According to A. Zahoor, in the 11th century, the Persian polymath scholar Abu Rayhan Biruni described the Milky Way galaxy as a multitude of fragments having the properties of nebulous stars, and also gave the latitudes of various stars during a lunar eclipse in 1019.

According to Josep Puig, the Andalusian astronomer Ibn Bajjah proposed that the Milky Way was made up of many stars that almost touched one another and appeared to be a continuous image due to the effect of refraction from sublunary material, citing his observation of the conjunction of Jupiter and Mars on 500 AH (1106/1107 AD) as evidence. Early European astronomers such as Tycho Brahe identified new stars in the night sky (later termed *novae*), suggesting that the heavens were not immutable. In 1584 Giordano Bruno suggested that the stars were like the Sun, and may have other planets, possibly even Earth-like, in orbit around them, an idea that had been suggested earlier by the ancient Greek philosophers, Democritus and Epicurus, and by medieval Islamic cosmologists such as Fakhr al-Din al-Razi. By the following century, the idea of the stars being the same as the Sun was reaching a consensus among astronomers. To explain why these stars exerted no net gravitational pull on the Solar System, Isaac Newton suggested that the stars were equally distributed in every direction, an idea prompted by the theologian Richard Bentley.

The Italian astronomer Geminiano Montanari recorded observing variations in luminosity of the star Algol in 1667. Edmond Halley published the first measurements of the proper motion of a pair of nearby "fixed" stars, demonstrating that they had changed positions since the time of the ancient Greek astronomers Ptolemy and Hipparchus.

William Herschel was the first astronomer to attempt to determine the distribution of stars in the sky. During the 1780s he established a series of gauges in 600 directions and counted the stars observed along each line of sight. From this he deduced that the number of stars steadily increased toward one side of the sky, in the direction of the Milky Way core. His son John Herschel repeated this study in the southern hemisphere and found a corresponding increase in the same direction. In addition to his other accomplishments, William Herschel is also noted for his discovery that some stars do not merely lie along the same line of sight, but are also physical companions that form binary star systems.

The science of stellar spectroscopy was pioneered by Joseph von Fraunhofer and Angelo Secchi. By comparing the spectra of stars such as Sirius to the Sun, they found differences in the strength and number of their absorption lines—the dark lines in a stellar spectra caused by the atmosphere's absorption of specific frequencies. In 1865 Secchi began classifying stars into spectral types. However, the modern version of the stellar classification scheme was developed by Annie J. Cannon during the 1900s.

Alpha Centauri A and B over limb of Saturn.

The first direct measurement of the distance to a star (61 Cygni at 11.4 light-years) was made in 1838 by Friedrich Bessel using the parallax technique. Parallax measurements demonstrated the vast separation of the stars in the heavens. Observation of double stars gained increasing importance during the 19th century. In 1834, Friedrich Bessel observed changes in the proper motion of the star Sirius and inferred a hidden companion. Edward Pickering discovered the first spectroscopic binary in 1899 when he observed the periodic splitting of the spectral lines of the star Mizar in a 104-day period. Detailed observations of many binary star systems were collected by astronomers such as William Struve and S. W. Burnham, allowing the masses of stars to be determined from computation of orbital elements. The first solution to the problem of deriving an orbit of binary stars from telescope observations was made by Felix Savary in 1827. The twentieth century saw increasingly rapid advances in the scientific study of stars. The photograph became a valuable astronomical tool. Karl Schwarzschild discovered that the color of a star and, hence, its temperature, could be determined by comparing the visual magnitude against the photographic magnitude. The development of the photoelectric photometer allowed precise measurements of magnitude at multiple wavelength intervals. In 1921 Albert A. Michelson made the first measurements of a stellar diameter using an interferometer on the Hooker telescope at Mount Wilson Observatory.

Important theoretical work on the physical structure of stars occurred during the first decades of the twentieth century. In 1913, the Hertzsprung-Russell diagram was developed, propelling the astrophysical study of stars. Successful models were developed to explain the interiors of stars and stellar evolution. Cecilia Payne-Gaposchkin first proposed that stars were made primarily of hydrogen and helium in her 1925 PhD thesis. The spectra of stars were further understood through advances in quantum physics. This allowed the chemical composition of the stellar atmosphere to be determined.

With the exception of supernovae, individual stars have primarily been observed in the Local Group, and especially in the visible part of the Milky Way (as demonstrated by the detailed star catalogues available for our galaxy). But some stars have been observed in the M100 galaxy of the Virgo Cluster, about 100 million light years from the Earth. In the Local Supercluster it is possible to see star clusters, and current telescopes could in principle observe faint individual stars in the Local Group. However, outside the Local Supercluster of galaxies, neither individual stars nor clusters of stars have been observed. The only exception is a faint image of a large star cluster containing hundreds of thousands of stars located at a distance of one billion light years—ten times further than the most distant star cluster previously observed.

Designations

This view contains blue stars known as "Blue stragglers", for their apparent location on the Hertzsprung–Russell diagram.

The concept of a constellation was known to exist during the Babylonian period. Ancient sky watchers imagined that prominent arrangements of stars formed patterns, and they associated these with particular aspects of nature or their myths. Twelve of these formations lay along the band of the ecliptic and these became the basis of astrology. Many of the more prominent individual stars were also given names, particularly with Arabic or Latin designations.

As well as certain constellations and the Sun itself, individual stars have their own myths. To the represented various important deities, from which the names of the planets Mercury, Venus, Mars, Jupiter and Saturn were taken. (Uranus and Neptune were also Greek and Roman gods, but neither planet was known in Antiquity because of their low brightness. Their names were assigned by later astronomers.)

Circa 1600, the names of the constellations were used to name the stars in the corresponding regions of the sky. The German astronomer Johann Bayer created a series of star maps and applied Greek letters as designations to the stars in each constellation. Later a numbering system based on the star's right ascension was invented and added to John Flamsteed's star catalogue in his book *"Historia coelestis Britannica"* (the 1712 edition), whereby this numbering system came to be called *Flamsteed designation* or *Flamsteed numbering*.

The only internationally recognized authority for naming celestial bodies is the International Astronomical Union (IAU). A number of private companies sell names of stars, which the British Library calls an unregulated commercial enterprise. The IAU has disassociated itself from this commercial practice, and these names are neither recognized by the IAU nor used by them. One such star-naming company is the International Star Registry, which, during the 1980s, was accused of deceptive practice for making it appear that the assigned name was official. This now-discontinued ISR practice was informally labeled a scam and a fraud, and the New York City Department of Consumer Affairs issued a violation against ISR for engaging in a deceptive trade practice.

Units of Measurement

Although stellar parameters can be expressed in SI units or CGS units, it is often most convenient to express mass, luminosity, and radii in solar units, based on the characteristics of the Sun:

solar mass:	$M_{\square} = 1.9891 \times 10^{30}$ kg
solar luminosity:	$L_{\square} = 3.827 \times 10^{26}$ W
solar radius	$R_{\square} = 6.960 \times 10^{8}$ m

Large lengths, such as the radius of a giant star or the semi-major axis of a binary star system, are often expressed in terms of the astronomical unit —approximately equal to the mean distance between the Earth and the Sun (150 million km or 93 million miles).

Formation and Evolution

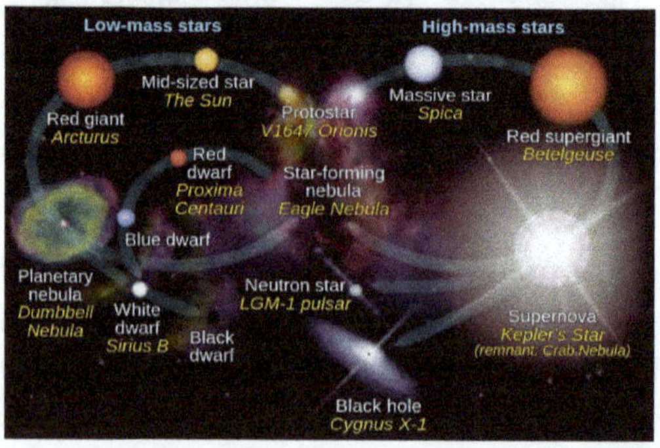

Stellar evolution of low-mass (left cycle) and high-mass (right cycle) stars, with examples in italics.

Stars condense from regions of space of higher density, yet those regions are less dense than within a vacuum chamber. These regions - known as *molecular clouds* - consist mostly of hydrogen, with about 23 to 28 percent helium and a few percent heavier elements. One example of such a star-forming region is the Orion Nebula. Most stars form in groups of dozens to hundreds of thousands of stars. Massive stars in these groups may powerfully illuminate those clouds, ionizing the hydrogen, and creating H II regions. Such feedback effects, from star formation, may ultimately disrupt the cloud and prevent further star formation.

All stars spend the majority of their existence as *main sequence stars*, fueled primarily by the nuclear fusion of hydrogen into helium within their cores. However, stars of different masses have markedly different properties at various stages of their development. The ultimate fate of more massive stars differs from that of less massive stars, as do their luminosities and the impact they have on their environment. Accordingly, astronomers often group stars by their mass:

Very low mass stars, with masses below 0.5 $M_{\odot}$, are fully convective and distribute helium evenly throughout the whole star while on the main sequence. Therefore, they never undergo shell burning, never become red giants, which cease fusing and become helium white dwarfs and slowly cool after exhausting their hydrogen. However, as the lifetime of 0.5 $M_{\odot}$ stars is longer than the age of the universe, no such star has yet reached the white dwarf stage.

Low mass stars (including the Sun), with a mass between 0.5 $M_{\odot}$ and 1.8–2.5 $M_{\odot}$ depending on composition, do become red giants as their core hydrogen is depleted and they begin to burn helium in core in a helium flash; they develop a degenerate carbon-oxygen core later on the asymptotic giant branch; they finally blow off their outer shell as a planetary nebula and leave behind their core in the form of a white dwarf.

Intermediate-mass stars, between 1.8–2.5 $M_{\odot}$ and 5–10 $M_{\odot}$, pass through evolutionary stages similar to low mass stars, but after a relatively short period on the RGB they ignite helium without a flash and spend an extended period in the red clump before forming a degenerate carbon-oxygen core.

Massive stars generally have a minimum mass of 7–10 $M_{\odot}$ (possibly as low as 5–6 $M_{\odot}$). After exhausting the hydrogen at the core these stars become supergiants and go on to fuse elements heavier than helium. They end their lives when their cores collapse and they explode as supernovae.

Star Formation

The formation of a star begins with gravitational instability within a molecular cloud, caused by regions of higher density - often triggered by compression of clouds by radiation from massive stars, expanding bubbles in the interstellar medium, the collision of different molecular clouds, or the collision of galaxies (as in a starburst galaxy). When a region reaches a sufficient density of matter to satisfy the criteria for Jeans instability, it begins to collapse under its own gravitational force.

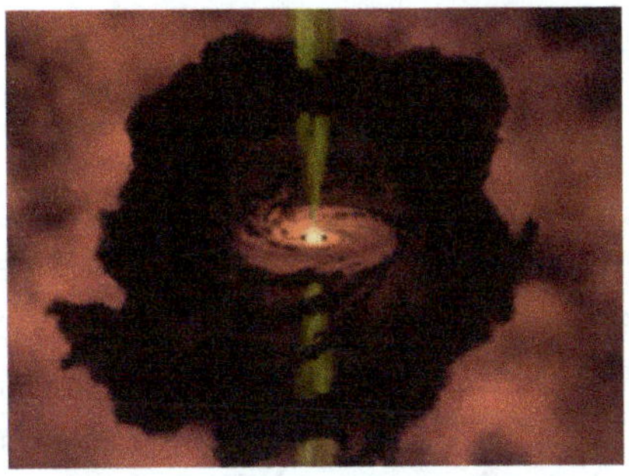

Artist's conception of the birth of a star within a dense molecular cloud.

As the cloud collapses, individual conglomerations of dense dust and gas form "Bok globules". As a globule collapses and the density increases, the gravitational energy converts into heat and the temperature rises. When the protostellar cloud has approximately reached the stable condition of hydrostatic equilibrium, a protostar forms at the core. These pre–main sequence stars are often surrounded by a protoplanetary disk and powered mainly by the conversion of gravitational energy. The period of gravitational contraction lasts about 10 to 15 million years.

A cluster of approximately 500 young stars lies within the nearby W40 stellar nursery.

Early stars of less than 2 $M_{\odot}$ are called T Tauri stars, while those with greater mass are Herbig Ae/Be stars. These newly formed stars emit jets of gas along their axis of rotation, which may reduce the angular momentum of the collapsing star and result in small patches of nebulosity known as Herbig–Haro objects. These jets, in combination with radiation from nearby massive stars, may help to drive away the surrounding cloud from which the star was formed.

Early in their development, T Tauri stars follow the Hayashi track—they contract and decrease in luminosity while remaining at roughly the same temperature. Less massive T Tauri stars follow this track to the main sequence, while more massive stars turn onto the Henyey track.

Most stars are observed to be members of binary star systems, and the properties of those binaries are the result of the conditions in which they formed. A gas cloud must lose its angular momentum in order to collapse and form a star. The fragmentation of the cloud into multiple stars distributes some of that angular momentum. The primordial binaries transfer some angular momentum by gravitational interactions during close encounters with other stars in young stellar clusters. These interactions tend to split apart more widely separated (soft) binaries while causing hard binaries to become more tightly bound. This produces the separation of binaries into their two observed populations distributions.

Main Sequence

Stars spend about 90% of their existence fusing hydrogen into helium in high-temperature and high-pressure reactions near the core. Such stars are said to be on the main sequence, and are called dwarf stars. Starting at zero-age main sequence, the proportion of helium in a star's core will steadily increase, the rate of nuclear fusion at the core will slowly increase, as will the star's temperature and luminosity. The Sun, for example, is estimated to have increased in luminosity by about 40% since it reached the main sequence 4.6 billion (4.6×10^9) years ago.

Every star generates a stellar wind of particles that causes a continual outflow of gas into space. For most stars, the mass lost is negligible. The Sun loses $10^{-14} M_{\odot}$ every year, or about 0.01% of its total mass over its entire lifespan. However, very massive stars can lose 10^{-7} to $10^{-5} M_{\odot}$ each year, significantly affecting their evolution. Stars that begin with more than 50 $M_{\odot}$ can lose over half their total mass while on the main sequence.

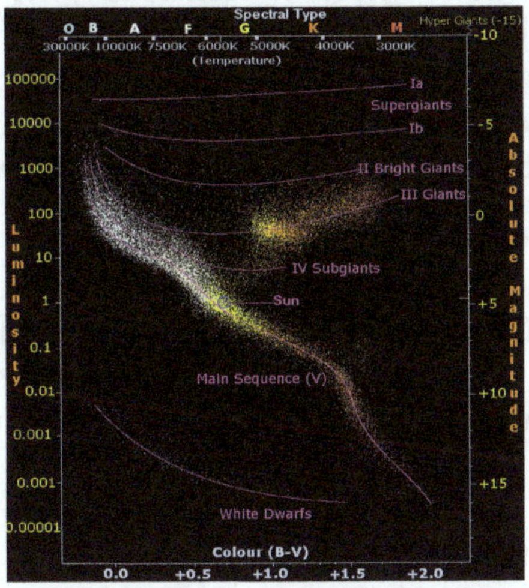

An example of a Hertzsprung–Russell diagram for a set of stars that includes the Sun (center).

The time a star spends on the main sequence depends primarily on the amount of fuel it has and the rate at which it fuses it. The Sun's is expected to live 10 billion (10^{10}) years. Massive stars consume their fuel very rapidly and are short-lived. Low mass stars consume their fuel very slowly. Stars less massive than 0.25 $M_{\odot}$, called red dwarfs, are able to fuse nearly all of their mass while stars of about 1 $M_{\odot}$ can only fuse about 10% of their mass. The combination of their slow fuel-consumption and relatively large usable fuel supply allows low mass stars to last about one trillion (10^{12}) years; the most extreme of 0.08 $M_{\odot}$) will last for about 12 trillion years. Red dwarfs become hotter and more luminous as they accumulate helium. When they eventually run out of hydrogen, they contract into a white dwarf and decline in temperature. However, since the lifespan of such stars is greater than the current age of the universe (13.8 billion years), no stars under about 0.85 $M_{\odot}$ are expected to have moved off the main sequence.

Besides mass, the elements heavier than helium can play a significant role in the evolution of stars. Astronomers label all elements heavier than helium "metals", and call the chemical concentration of these elements in a star, its metallicity. A star's metallicity can influence the time the star takes to burn its fuel, and controls the formation of its magnetic fields, which affects the strength of its stellar wind. Older, population II stars have substantially less metallicity than the younger, population I stars due to the composition of the molecular clouds from which they formed. Over time, such clouds become increasingly enriched in heavier elements as older stars die and shed portions of their atmospheres.

Post–Main Sequence

As stars of at least 0.4 $M_{\odot}$ exhaust their supply of hydrogen at their core, they start to fuse hydrogen in a shell outside the helium core. Their outer layers expand and cool greatly as they form a red giant. In about 5 billion years, when the Sun enters the helium burning phase, it will expand to a maximum radius of roughly 1 astronomical unit (150 million kilometres), 250 times its present size, and lose 30% of its current mass.

As the hydrogen shell burning produces more helium, the core increases in mass and temperature. In a red giant of up to 2.25 $M_{\odot}$, the mass of the helium core becomes degenerate prior to helium fusion. Finally, when the temperature increases sufficiently, helium fusion begins explosively in what is called a helium flash, and the star rapidly shrinks in radius, increases its surface temperature, and moves to the horizontal branch of the HR diagram. For more massive stars, helium core fusion starts before the core becomes degenerate, and the star spends some time in the red clump, slowly burning helium, before the outer convective envelope collapses and the star then moves to the horizontal branch.

After the star has fused the helium of its core, the carbon product fuses producing a hot core with an outer shell of fusing helium. The star then follows an evolutionary path called the asymptotic giant branch (AGB) that parallels the other described red giant phase, but with a higher luminosity. The more massive AGB stars may undergo a brief period of carbon fusion before the core becomes degenerate.

Massive Stars

During their helium-burning phase, stars of more than nine solar masses expand to form red supergiants. When this fuel is exhausted at the core, they continue to fuse elements heavier than helium.

The core contracts and the temperature and pressure rises enough to fuse carbon. This process continues, with the successive stages being fueled by neon, oxygen and silicon. Near the end of the star's life, fusion continues along a series of onion-layer shells within a massive star. Each shell fuses a different element, with the outermost shell fusing hydrogen; the next shell fusing helium, and so forth.

The final stage occurs when a massive star begins producing iron. Since iron nuclei are more tightly bound than any heavier nuclei, any fusion beyond iron does not produce a net release of energy. To a very limited degree such a process proceeds, but it consumes energy. Likewise, since they are more tightly bound than all lighter nuclei, such energy cannot be released by fission. In relatively old, very massive stars, a large core of inert iron will accumulate in the center of the star. The heavier elements in these stars can work their way to the surface, forming evolved objects known as Wolf-Rayet stars that have a dense stellar wind which sheds the outer atmosphere.

Collapse

As a star's core shrinks, the intensity of radiation from that surface increases, creating such radiation pressure on the outer shell of gas that it will push those layers away, forming a planetary nebula. If what remains after the outer atmosphere has been shed is less than 1.4 $M_{\odot}$, it shrinks to a relatively tiny object about the size of Earth, known as a white dwarf. White dwarfs lack the mass for further gravitational compression to take place. The electron-degenerate matter inside a white dwarf is no longer a plasma, even though stars are generally referred to as being spheres of plasma. Eventually, white dwarfs fade into black dwarfs over a very long period of time.

The Crab Nebula, remnants of a supernova that was first observed around 1050 AD.

In larger stars, fusion continues until the iron core has grown so large (more than 1.4 $M_\odot$) that it can no longer support its own mass. This core will suddenly collapse as its electrons are driven into its protons, forming neutrons, neutrinos, and gamma rays in a burst of electron capture and inverse beta decay. The shockwave formed by this sudden collapse causes the rest of the star to explode in a supernova. Supernovae become so bright that they may briefly outshine the star's entire home galaxy. When they occur within the Milky Way, supernovae have historically been observed by naked-eye observers as "new stars" where none seemingly existed before.

A supernova explosion blows away the star's outer layers, leaving a remnant such as the Crab Nebula. The core is compressed into a neutron star, which sometimes manifests itself as a pulsar or X-ray burster. In the case of the largest stars, the remnant is a black hole greater than 4 $M_\odot$s. In a neutron star the matter is in a state known as neutron-degenerate matter, with a more exotic form of degenerate matter, QCD matter, possibly present in the core. Within a black hole, the matter is in a state that is not currently understood.

The blown-off outer layers of dying stars include heavy elements, which may be recycled during the formation of new stars. These heavy elements allow the formation of rocky planets. The outflow from supernovae and the stellar wind of large stars play an important part in shaping the interstellar medium.

Binary Stars

The post–main-sequence evolution of binary stars may be significantly different from the evolution of single stars of the same mass. If stars in a binary system are sufficiently close, when one of the stars expands to become a red giant it may overflow its Roche lobe, the region around a star where material is gravitationally bound to that star, leading to transfer of material to the other. When the Roche lobe is violated, a variety of phenomena can result, including contact binaries, common-envelope binaries, cataclysmic variables, and type Ia supernovae.

Distribution

A white dwarf star in orbit around Sirius (artist's impression).

In addition to isolated stars, a multi-star system can consist of two or more gravitationally bound stars that orbit each other. The simplest and most common multi-star system is a binary star, but systems of three or more stars are also found. For reasons of orbital stability, such multi-star systems are often organized into hierarchical sets of binary stars. Larger groups called star clusters also exist. These range from loose stellar associations with only a few stars, up to enormous globular clusters with hundreds of thousands of stars. Such systems orbit our Milky Way galaxy.

It has been a long-held assumption that the majority of stars occur in gravitationally bound, multiple-star systems. This is particularly true for very massive O and B class stars, where 80% of the stars are believed to be part of multiple-star systems. The proportion of single star systems increases with decreasing star mass, so that only 25% of red dwarfs are known to have stellar companions. As 85% of all stars are red dwarfs, most stars in the Milky Way are likely single from birth.

Stars are not spread uniformly across the universe, but are normally grouped into galaxies along with interstellar gas and dust. A typical galaxy contains hundreds of billions of stars, and there are more than 100 billion (10^{11}) galaxies in the observable universe. In 2010, one estimate of the number of stars in the observable universe was 300 sextillion (3×10^{23}). While it is often believed that stars only exist within galaxies, intergalactic stars have been discovered.

The nearest star to the Earth, apart from the Sun, is Proxima Centauri, which is 39.9 trillion kilometres, or 4.2 light-years. Travelling at the orbital speed of the Space Shuttle (8 kilometres per second—almost 30,000 kilometres per hour), it would take about 150,000 years to arrive. This it typical of stellar separations in galactic discs. Stars can be much closer to each other in the centres of galaxies and in globular clusters, or much farther apart in galactic halos.

Due to the relatively vast distances between stars outside the galactic nucleus, collisions between stars are thought to be rare. In denser regions such as the core of globular clusters or the galactic center, collisions can be more common. Such collisions can produce what are known as blue stragglers. These abnormal stars have a higher surface temperature than the other main sequence stars with the same luminosity of the cluster to which it belongs.

Characteristics

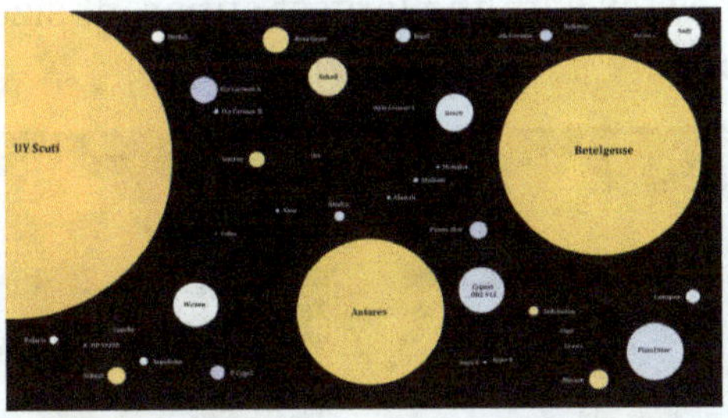

Some of the well-known stars with their apparent colors and relative sizes.

Almost everything about a star is determined by its initial mass, including such characteristics as luminosity, size, evolution, lifespan, and its eventual fate.

Age

Most stars are between 1 billion and 10 billion years old. Some stars may even be close to 13.8 billion years old—the observed age of the universe. The oldest star yet discovered, HD 140283, nicknamed Methuselah star, is an estimated 14.46 ± 0.8 billion years old. (Due to the uncertainty in the value, this age for the star does not conflict with the age of the Universe, determined by the Planck satellite as 13.799 ± 0.021).

The more massive the star, the shorter its lifespan, primarily because massive stars have greater pressure on their cores, causing them to burn hydrogen more rapidly. The most massive stars last an average of a few million years, while stars of minimum mass (red dwarfs) burn their fuel very slowly and can last tens to hundreds of billions of years.

Chemical Composition

When stars form in the present Milky Way galaxy they are composed of about 71% hydrogen and 27% helium, as measured by mass, with a small fraction of heavier elements. Typically the portion of heavy elements is measured in terms of the iron content of the stellar atmosphere, as iron is a common element and its absorption lines are relatively easy to measure. The portion of heavier elements may be an indicator of the likelihood that the star has a planetary system.

The star with the lowest iron content ever measured is the dwarf HE1327-2326, with only 1/200,000th the iron content of the Sun. By contrast, the super-metal-rich star μ Leonis has nearly double the abundance of iron as the Sun, while the planet-bearing star 14 Herculis has nearly triple the iron. There also exist chemically peculiar stars that show unusual abundances of certain elements in their spectrum; especially chromium and rare earth elements. Stars with cooler outer atmospheres, including the Sun, can form various diatomic and polyatomic molecules.

Diameter

Due to their great distance from the Earth, all stars except the Sun appear to the unaided eye as shining points in the night sky that twinkle because of the effect of the Earth's atmosphere. The

Sun is also a star, but it is close enough to the Earth to appear as a disk instead, and to provide daylight. Other than the Sun, the star with the largest apparent size is R Doradus, with an angular diameter of only 0.057 arcseconds.

Stars vary widely in size. In each image in the sequence, the right-most object appears as the left-most object in the next panel. The Earth appears at right in panel 1 and the Sun is second from the right in panel 3. The rightmost star at panel 6 is UY Scuti, the largest known star.

The disks of most stars are much too small in angular size to be observed with current ground-based optical telescopes, and so interferometer telescopes are required to produce images of these objects. Another technique for measuring the angular size of stars is through occultation. By precisely measuring the drop in brightness of a star as it is occulted by the Moon (or the rise in brightness when it reappears), the star's angular diameter can be computed.

Stars range in size from neutron stars, which vary anywhere from 20 to 40 km (25 mi) in diameter, to supergiants like Betelgeuse in the Orion constellation, which has a diameter approximately 1,070 times that of the Sun—about 1,490,171,880 km (925,949,878 mi). Betelgeuse, however, has a much lower density than the Sun.

Kinematics

The Pleiades, an open cluster of stars in the constellation of Taurus. These stars share a common motion through space.

The motion of a star relative to the Sun can provide useful information about the origin and age of a star, as well as the structure and evolution of the surrounding galaxy. The components of motion of a star consist of the radial velocity toward or away from the Sun, and the traverse angular movement, which is called its proper motion.

Radial velocity is measured by the doppler shift of the star's spectral lines, and is given in units of km/s. The proper motion of a star, its parallax, is determined by precise astrometric measurements in units of milli-arc seconds (mas) per year. With knowledge of the star's parallax and its distance, the proper motion velocity can be calculated. Together with the radial velocity, the total velocity can be calculated. Stars with high rates of proper motion are likely to be relatively close to the Sun, making them good candidates for parallax measurements.

When both rates of movement are known, the space velocity of the star relative to the Sun or the galaxy can be computed. Among nearby stars, it has been found that younger population I stars have generally lower velocities than older, population II stars. The latter have elliptical orbits that are inclined to the plane of the galaxy. A comparison of the kinematics of nearby stars has allowed astronomers to trace their origin to common points in giant molecular clouds, and are referred to as stellar associations.

Magnetic Field

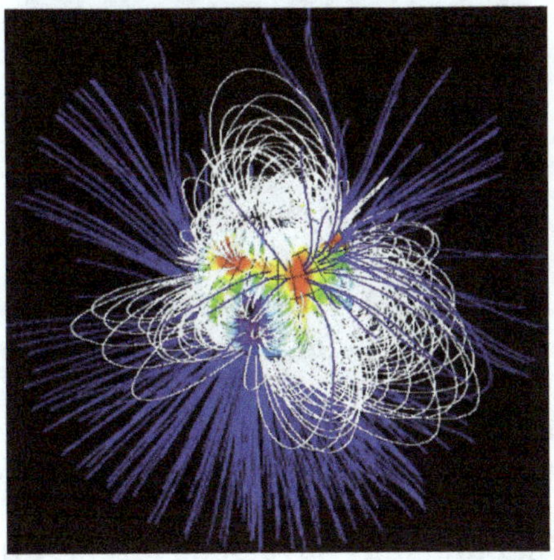

Surface magnetic field of SU Aur (a young star of T Tauri type), reconstructed by means of Zeeman-Doppler imaging.

The magnetic field of a star is generated within regions of the interior where convective circulation occurs. This movement of conductive plasma functions like a dynamo, wherein the movement of electrical charges induce magnetic fields, as does a mechanical dynamo. Those magnetic fields have a great range that extend throughout and beyond the star. The strength of the magnetic field varies with the mass and composition of the star, and the amount of magnetic surface activity depends upon the star's rate of rotation. This surface activity produces starspots, which are regions of strong magnetic fields and lower than normal surface temperatures. Coronal loops are arching magnetic field flux lines that rise from a star's surface into the star's outer atmosphere, its corona. The coronal loops can be seen due to the plasma they conduct along their length. Stellar flares are bursts of high-energy particles that are emitted due to the same magnetic activity.

Young, rapidly rotating stars tend to have high levels of surface activity because of their magnetic field. The magnetic field can act upon a star's stellar wind, functioning as a brake to gradually slow the rate of rotation with time. Thus, older stars such as the Sun have a much slower rate of rotation and a lower level of surface activity. The activity levels of slowly rotating stars tend to vary in a cyclical manner and can shut down altogether for periods of time. During the Maunder minimum, for example, the Sun underwent a 70-year period with almost no sunspot activity.

Mass

One of the most massive stars known is Eta Carinae, which, with 100–150 times as much mass as the Sun, will have a lifespan of only several million years. Studies of the most massive open clusters suggests 150 $M_{\odot}$ as an upper limit for stars in the current era of the universe. This represents an empirical value for the theoretical limit on the mass of forming stars due to increasing radiation pressure on the accreting gas cloud. Several stars in the R136 cluster in the Large Magellanic Cloud have been measured with larger masses, but it has been determined that they could have been created through the collision and merger of massive stars in close binary systems, sidestepping the 150 $M_{\odot}$ limit on massive star formation.

The reflection nebula NGC 1999 is brilliantly illuminated by V380 Orionis (center), a variable star with about 3.5 times the mass of the Sun. The black patch of sky is a vast hole of empty space and not a dark nebula as previously thought.

The first stars to form after the Big Bang may have been larger, up to 300 $M_{\odot}$, due to the complete absence of elements heavier than lithium in their composition. This generation of supermassive population III stars is likely to have existed in the very early universe (i.e., they are observed to have a high redshift), and may have started the production of chemical elements heavier than hydrogen that are needed for the later formation of planets and life. In June 2015, astronomers reported evidence for Population III stars in the Cosmos Redshift 7 galaxy at $z = 6.60$.

With a mass only 80 times that of Jupiter (M_J), 2MASS J0523-1403 is the smallest known star undergoing nuclear fusion in its core. For stars with metallicity similar to the Sun, the theoretical minimum mass the star can have and still undergo fusion at the core, is estimated to be about 75 M_J. When the metallicity is very low, however, the minimum star size seems to be about 8.3% of the solar mass, or about 87 M_J. Smaller bodies called brown dwarfs, occupy a poorly defined grey area between stars and gas giants.

The combination of the radius and the mass of a star determines its surface gravity. Giant stars have a much lower surface gravity than do main sequence stars, while the opposite is the case for degenerate, compact stars such as white dwarfs. The surface gravity can influence the appearance of a star's spectrum, with higher gravity causing a broadening of the absorption lines.

Rotation

The rotation rate of stars can be determined through spectroscopic measurement, or more exactly determined by tracking their starspots. Young stars can have a rotation greater than

100 km/s at the equator. The B-class star Achernar, for example, has an equatorial velocity of about 225 km/s or greater, causing its equator to be slung outward and giving it an equatorial diameter that is more than 50% greater than between the poles. This rate of rotation is just below the critical velocity of 300 km/s at which speed the star would break apart. By contrast, the Sun rotates once every 25 – 35 days, with an equatorial velocity of 1.994 km/s. A main sequence star's magnetic field and the stellar wind serve to slow its rotation by a significant amount as it evolves on the main sequence.

Degenerate stars have contracted into a compact mass, resulting in a rapid rate of rotation. However they have relatively low rates of rotation compared to what would be expected by conservation of angular momentum—the tendency of a rotating body to compensate for a contraction in size by increasing its rate of spin. A large portion of the star's angular momentum is dissipated as a result of mass loss through the stellar wind. In spite of this, the rate of rotation for a pulsar can be very rapid. The pulsar at the heart of the Crab nebula, for example, rotates 30 times per second. The rotation rate of the pulsar will gradually slow due to the emission of radiation.

Temperature

The surface temperature of a main sequence star is determined by the rate of energy production of its core and by its radius, and is often estimated from the star's color index. The temperature is normally given in terms of an effective temperature, which is the temperature of an idealized black body that radiates its energy at the same luminosity per surface area as the star. Note that the effective temperature is only a representative of the surface, as the temperature increases toward the core. The temperature in the core region of a star is several million kelvins.

The stellar temperature will determine the rate of ionization of various elements, resulting in characteristic absorption lines in the spectrum. The surface temperature of a star, along with its visual absolute magnitude and absorption features, is used to classify a star.

Massive main sequence stars can have surface temperatures of 50,000 K. Smaller stars such as the Sun have surface temperatures of a few thousand K. Red giants have relatively low surface temperatures of about 3,600 K; but they also have a high luminosity due to their large exterior surface area.

Radiation

The energy produced by stars, a product of nuclear fusion, radiates to space as both electromagnetic radiation and particle radiation. The particle radiation emitted by a star is manifested as the stellar wind, which streams from the outer layers as electrically charged protons and alpha and beta particles. Although almost massless, there also exists a steady stream of neutrinos emanating from the star's core.

The production of energy at the core is the reason stars shine so brightly: every time two or more atomic nuclei fuse together to form a single atomic nucleus of a new heavier element, gamma ray photons are released from the nuclear fusion product. This energy is converted to other forms of electromagnetic energy of lower frequency, such as visible light, by the time it reaches the star's outer layers.

The color of a star, as determined by the most intense frequency of the visible light, depends on the temperature of the star's outer layers, including its photosphere. Besides visible light, stars also emit forms of electromagnetic radiation that are invisible to the human eye. In fact, stellar electromagnetic radiation spans the entire electromagnetic spectrum, from the longest wavelengths of radio waves through infrared, visible light, ultraviolet, to the shortest of X-rays, and gamma rays. From the standpoint of total energy emitted by a star, not all components of stellar electromagnetic radiation are significant, but all frequencies provide insight into the star's physics.

Using the stellar spectrum, astronomers can also determine the surface temperature, surface gravity, metallicity and rotational velocity of a star. If the distance of the star is found, such as by measuring the parallax, then the luminosity of the star can be derived. The mass, radius, surface gravity, and rotation period can then be estimated based on stellar models. (Mass can be calculated for stars in binary systems by measuring their orbital velocities and distances. Gravitational microlensing has been used to measure the mass of a single star.) With these parameters, astronomers can also estimate the age of the star.

Luminosity

The luminosity of a star is the amount of light and other forms of radiant energy it radiates per unit of time. It has units of power. The luminosity of a star is determined by its radius and surface temperature. Many stars do not radiate uniformly across their entire surface. The rapidly rotating star Vega, for example, has a higher energy flux (power per unit area) at its poles than along its equator.

Patches of the star's surface with a lower temperature and luminosity than average are known as starspots. Small, *dwarf* stars such as our Sun generally have essentially featureless disks with only small starspots. *Giant* stars have much larger, more obvious starspots, and they also exhibit strong stellar limb darkening. That is, the brightness decreases towards the edge of the stellar disk. Red dwarf flare stars such as UV Ceti may also possess prominent starspot features.

Magnitude

The apparent brightness of a star is expressed in terms of its apparent magnitude. It is a function of the star's luminosity, its distance from Earth, and the altering of the star's light as it passes through Earth's atmosphere. Intrinsic or absolute magnitude is directly related to a star's luminosity, and is what the apparent magnitude a star would be if the distance between the Earth and the star were 10 parsecs (32.6 light-years).

Number of stars brighter than magnitude	
Apparent magnitude	Number of stars
0	4
1	15
2	48
3	171
4	513

5	1,602
6	4,800
7	14,000

Both the apparent and absolute magnitude scales are logarithmic units: one whole number difference in magnitude is equal to a brightness variation of about 2.5 times (the 5th root of 100 or approximately 2.512). This means that a first magnitude star (+1.00) is about 2.5 times brighter than a second magnitude (+2.00) star, and about 100 times brighter than a sixth magnitude star (+6.00). The faintest stars visible to the naked eye under good seeing conditions are about magnitude +6.

On both apparent and absolute magnitude scales, the smaller the magnitude number, the brighter the star; the larger the magnitude number, the fainter the star. The brightest stars, on either scale, have negative magnitude numbers. The variation in brightness (ΔL) between two stars is calculated by subtracting the magnitude number of the brighter star (m_b) from the magnitude number of the fainter star (m_f), then using the difference as an exponent for the base number 2.512; that is to say:

$$\Delta m = m_f - m_b$$

$$2.512^{\Delta m} = \Delta L$$

Relative to both luminosity and distance from Earth, a star's absolute magnitude (M) and apparent magnitude (m) are not equivalent; for example, the bright star Sirius has an apparent magnitude of −1.44, but it has an absolute magnitude of +1.41.

The Sun has an apparent magnitude of −26.7, but its absolute magnitude is only +4.83. Sirius, the brightest star in the night sky as seen from Earth, is approximately 23 times more luminous than the Sun, while Canopus, the second brightest star in the night sky with an absolute magnitude of −5.53, is approximately 14,000 times more luminous than the Sun. Despite Canopus being vastly more luminous than Sirius, however, Sirius appears brighter than Canopus. This is because Sirius is merely 8.6 light-years from the Earth, while Canopus is much farther away at a distance of 310 light-years.

As of 2006, the star with the highest known absolute magnitude is LBV 1806-20, with a magnitude of −14.2. This star is at least 5,000,000 times more luminous than the Sun. The least luminous stars that are currently known are located in the NGC 6397 cluster. The faintest red dwarfs in the cluster were magnitude 26, while a 28th magnitude white dwarf was also discovered. These faint stars are so dim that their light is as bright as a birthday candle on the Moon when viewed from the Earth.

Classification

Surface temperature ranges for different stellar classes		
Class	**Temperature**	**Sample star**
O	33,000 K or more	Zeta Ophiuchi
B	10,500–30,000 K	Rigel
A	7,500–10,000 K	Altair

F	6,000–7,200 K	Procyon A
G	5,500–6,000 K	Sun
K	4,000–5,250 K	Epsilon Indi
M	2,600–3,850 K	Proxima Centauri

The current stellar classification system originated in the early 20th century, when stars were classified from *A* to *Q* based on the strength of the hydrogen line. It thought that the hydrogen line strength was a simple linear function of temperature. Rather, it was more complicated; it strengthened with increasing temperature, it peaked near 9000 K, and then declined at greater temperatures. When the classifications were reordered by temperature, it more closely resembled the modern scheme.

Stars are given a single-letter classification according to their spectra, ranging from type *O*, which are very hot, to *M*, which are so cool that molecules may form in their atmospheres. The main classifications in order of decreasing surface temperature are: *O, B, A, F, G, K,* and *M*. A variety of rare spectral types are given special classifications. The most common of these are types *L* and *T*, which classify the coldest low-mass stars and brown dwarfs. Each letter has 10 sub-divisions, numbered from 0 to 9, in order of decreasing temperature. However, this system breaks down at extreme high temperatures as classes *O0* and *O1* may not exist.

In addition, stars may be classified by the luminosity effects found in their spectral lines, which correspond to their spatial size and is determined by their surface gravity. These range from *0* (hypergiants) through *III* (giants) to *V* (main sequence dwarfs); some authors add *VII* (white dwarfs). Most stars belong to the main sequence, which consists of ordinary hydrogen-burning stars. These fall along a narrow, diagonal band when graphed according to their absolute magnitude and spectral type. The Sun is a main sequence *G2V* yellow dwarf of intermediate temperature and ordinary size.

Additional nomenclature, in the form of lower-case letters added to the end of the spectral type to indicate peculiar features of the spectrum. For example, an "*e*" can indicate the presence of emission lines; "*m*" represents unusually strong levels of metals, and "*var*" can mean variations in the spectral type.

White dwarf stars have their own class that begins with the letter *D*. This is further sub-divided into the classes *DA, DB, DC, DO, DZ,* and *DQ*, depending on the types of prominent lines found in the spectrum. This is followed by a numerical value that indicates the temperature.

Variable Stars

Variable stars have periodic or random changes in luminosity because of intrinsic or extrinsic properties. Of the intrinsically variable stars, the primary types can be subdivided into three principal groups.

During their stellar evolution, some stars pass through phases where they can become pulsating variables. Pulsating variable stars vary in radius and luminosity over time, expanding and contracting with periods ranging from minutes to years, depending on the size of the star. This category includes Cepheid and Cepheid-like stars, and long-period variables such as Mira.

The asymmetrical appearance of Mira, an oscillating variable star.

Eruptive variables are stars that experience sudden increases in luminosity because of flares or mass ejection events. This group includes protostars, Wolf-Rayet stars, and flare stars, as well as giant and supergiant stars.

Cataclysmic or explosive variable stars are those that undergo a dramatic change in their properties. This group includes novae and supernovae. A binary star system that includes a nearby white dwarf can produce certain types of these spectacular stellar explosions, including the nova and a Type 1a supernova. The explosion is created when the white dwarf accretes hydrogen from the companion star, building up mass until the hydrogen undergoes fusion. Some novae are also recurrent, having periodic outbursts of moderate amplitude.

Stars can also vary in luminosity because of extrinsic factors, such as eclipsing binaries, as well as rotating stars that produce extreme starspots. A notable example of an eclipsing binary is Algol, which regularly varies in magnitude from 2.3 to 3.5 over a period of 2.87 days.

Structure

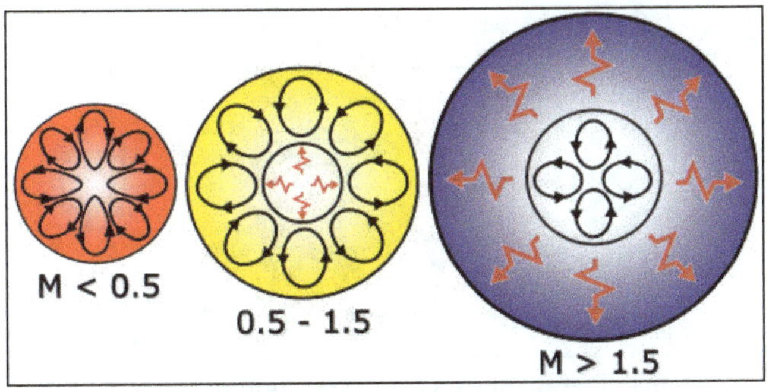

Internal structures of main sequence stars, convection zones with arrowed cycles and radiative zones with red flashes. To the left a **low-mass** red dwarf, in the center a **mid-sized** yellow dwarf, and, at the right, a **massive** blue-white main sequence star.

The interior of a stable star is in a state of hydrostatic equilibrium: the forces on any small volume almost exactly counterbalance each other. The balanced forces are inward gravitational force and an outward force due to the pressure gradient within the star. The pressure gradient is established by the temperature gradient of the plasma; the outer part of the star is cooler than the core. The temperature at the core of a main sequence or giant star is at least on the order of 10^7 K. The resulting temperature and pressure at the hydrogen-burning core of a main sequence star are sufficient for nuclear fusion to occur and for sufficient energy to be produced to prevent further collapse of the star.

As atomic nuclei are fused in the core, they emit energy in the form of gamma rays. These photons interact with the surrounding plasma, adding to the thermal energy at the core. Stars on the main sequence convert hydrogen into helium, creating a slowly but steadily increasing proportion of helium in the core. Eventually the helium content becomes predominant, and energy production ceases at the core. Instead, for stars of more than 0.4 $M_{\odot}$, fusion occurs in a slowly expanding shell around the degenerate helium core.

In addition to hydrostatic equilibrium, the interior of a stable star will also maintain an energy balance of thermal equilibrium. There is a radial temperature gradient throughout the interior that results in a flux of energy flowing toward the exterior. The outgoing flux of energy leaving any layer within the star will exactly match the incoming flux from below.

The radiation zone is the region of the stellar interior where the flux of energy outward is dependent on radiative heat transfer, since convective heat transfer is inefficient in that zone. In this region the plasma will not be perturbed, and any mass motions will die out. If this is not the case, however, then the plasma becomes unstable and convection will occur, forming a convection zone. This can occur, for example, in regions where very high energy fluxes occur, such as near the core or in areas with high opacity (making radiatative heat transfer inefficient) as in the outer envelope.

The occurrence of convection in the outer envelope of a main sequence star depends on the star's mass. Stars with several times the mass of the Sun have a convection zone deep within the interior and a radiative zone in the outer layers. Smaller stars such as the Sun are just the opposite, with the convective zone located in the outer layers. Red dwarf stars with less than 0.4 $M_{\odot}$ are convective throughout, which prevents the accumulation of a helium core. For most stars the convective zones will also vary over time as the star ages and the constitution of the interior is modified.

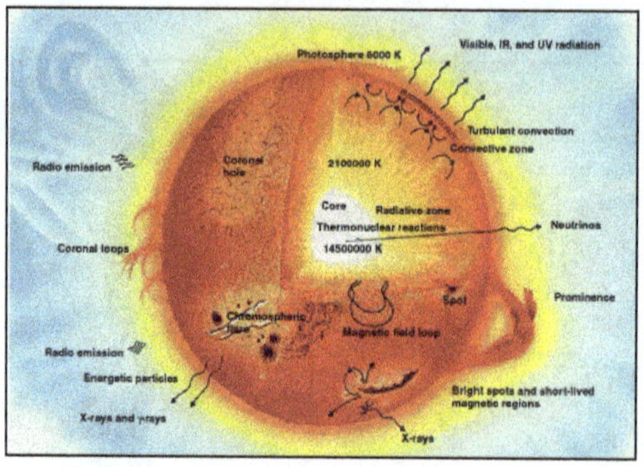

This diagram shows a cross-section of the Sun.

The photosphere is that portion of a star that is visible to an observer. This is the layer at which the plasma of the star becomes transparent to photons of light. From here, the energy generated at the core becomes free to propagate into space. It is within the photosphere that sun spots, regions of lower than average temperature, appear.

Above the level of the photosphere is the stellar atmosphere. In a main sequence star such as the Sun, the lowest level of the atmosphere, just above the photosphere, is the thin chromosphere region, where spicules appear and stellar flares begin. Above this is the transition region, where the temperature rapidly increases within a distance of only 100 km (62 mi). Beyond this is the corona, a volume of super-heated plasma that can extend outward to several million kilometres. The existence of a corona appears to be dependent on a convective zone in the outer layers of the star. Despite its high temperature, and the corona emits very little light, due to its low gas density. The corona region of the Sun is normally only visible during a solar eclipse.

From the corona, a stellar wind of plasma particles expands outward from the star, until it interacts with the interstellar medium. For the Sun, the influence of its solar wind extends throughout a bubble-shaped region called the heliosphere.

Nuclear Fusion Reaction Pathways

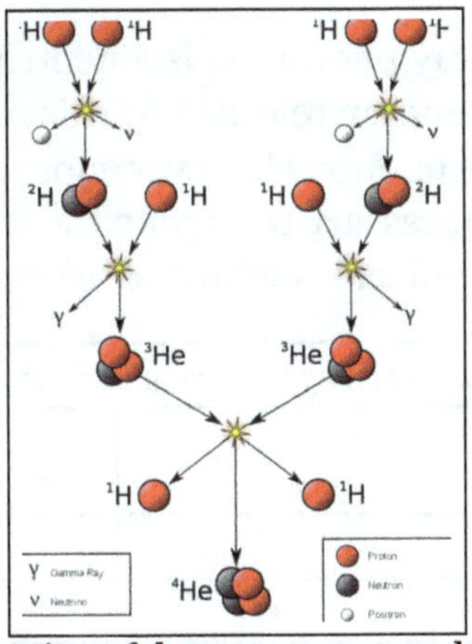

Overview of the proton-proton chain.

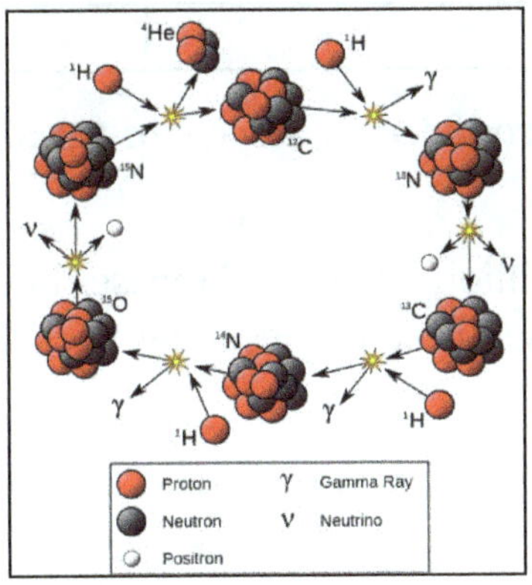

The carbon-nitrogen-oxygen cycle.

A variety of nuclear fusion reactions take place in the cores of stars, that depend upon their mass and composition. When nuclei fuse, the mass of the fused product is less than the mass of the original parts. This lost mass is converted to electromagnetic energy, according to the mass-energy equivalence relationship $E = mc^2$.

The hydrogen fusion process is temperature-sensitive, so a moderate increase in the core temperature will result in a significant increase in the fusion rate. As a result, the core temperature of main sequence stars only varies from 4 million kelvin for a small M-class star to 40 million kelvin for a massive O-class star.

In the Sun, with a 10-million-kelvin core, hydrogen fuses to form helium in the proton-proton chain reaction:

$$4^1H \rightarrow 2^2H + 2e^+ + 2\nu_e (2 \times 0.4 \text{ MeV})$$

$$2e^+ + 2e^- \rightarrow 2\gamma \ (2 \times 1.0 \text{ MeV})$$

$$2^1H + 2^2H \rightarrow 2^3He + 2\gamma \ (2 \times 5.5 \text{ MeV})$$

$$2^3He \rightarrow {}^4He + 2^1H \ (12.9 \text{ MeV})$$

These reactions result in the overall reaction:

$$4^1H \rightarrow {}^4He + 2e^+ + 2\gamma + 2\nu_e \ (26.7 \text{ MeV})$$

where e^+ is a positron, γ is a gamma ray photon, ν_e is a neutrino, and H and He are isotopes of hydrogen and helium, respectively. The energy released by this reaction is in millions of electron volts, which is actually only a tiny amount of energy. However enormous numbers of these reactions occur constantly, producing all the energy necessary to sustain the star's radiation output. In comparison, the combustion of two hydrogen gas molecules with one oxygen gas molecule releases only 5.7 eV.

Minimum stellar mass required for fusion	
Element	Solar masses
Hydrogen	0.01
Helium	0.4
Carbon	5
Neon	8

In more massive stars, helium is produced in a cycle of reactions catalyzed by carbon called the carbon-nitrogen-oxygen cycle.

In evolved stars with cores at 100 million kelvin and masses between 0.5 and 10 $M_\odot$, helium can be transformed into carbon in the triple-alpha process that uses the intermediate element beryllium:

$${}^4He + {}^4He + 92 \text{ keV} \rightarrow {}^{8*}Be$$

$${}^4He + {}^{8*}Be + 67 \text{ keV} \rightarrow {}^{12*}C$$

$${}^{12*}C \rightarrow {}^{12}C + \gamma + 7.4 \text{ MeV}$$

For an overall reaction of:

$$3^4He \rightarrow {}^{12}C + \gamma + 7.2 \text{ MeV}$$

In massive stars, heavier elements can also be burned in a contracting core through the neon burning process and oxygen burning process. The final stage in the stellar nucleosynthesis process is the silicon burning process that results in the production of the stable isotope iron-56, an endothermic process that consumes energy, and so further energy can only be produced through gravitational collapse.

The example below shows the amount of time required for a star of 20 $M_\odot$ to consume all of its nuclear fuel. As an O-class main sequence star, it would be 8 times the solar radius and 62,000 times the Sun's luminosity.

Fuel material	Temperature (million kelvins)	Density (kg/cm³)	Burn duration (τ in years)
H	37	0.0045	8.1 million
He	188	0.97	1.2 million
C	870	170	976
Ne	1,570	3,100	0.6
O	1,980	5,550	1.25
S/Si	3,340	33,400	0.0315

Star Formation

Star formation is the process by which dense regions within molecular clouds in interstellar space, sometimes referred to as "stellar nurseries" or "star-forming regions", fuse to form stars. As a branch of astronomy, star formation includes the study of the interstellar medium (ISM) and giant molecular clouds (GMC) as precursors to the star formation process, and the study of protostars and young stellar objects as its immediate products. It is closely related to planet formation, another branch of astronomy. Star formation theory, as well as accounting for the formation of a single star, must also account for the statistics of binary stars and the initial mass function.

In June 2015, astronomers reported evidence for Population III stars in the Cosmos Redshift 7 galaxy at $z = 6.60$. Such stars are likely to have existed in the very early universe (i.e., at high redshift), and may have started the production of chemical elements heavier than hydrogen that are needed for the later formation of planets and life as we know it.

Stellar Nurseries

Hubble telescope image known as *Pillars of Creation,* where stars are forming in the Eagle Nebula.

Interstellar Clouds

A spiral galaxy like the Milky Way contains stars, stellar remnants, and a diffuse interstellar medium (ISM) of gas and dust. The interstellar medium consists of 10^{-4} to 10^{6} particles per cm^3 and is typically composed of roughly 70% hydrogen by mass, with most of the remaining gas consisting of helium. This medium has been chemically enriched by trace amounts of heavier elements that were ejected from stars as they passed beyond the end of their main sequence lifetime. Higher density regions of the interstellar medium form clouds, or *diffuse nebulae*, where star formation takes place. In contrast to spirals, an elliptical galaxy loses the cold component of its interstellar medium within roughly a billion years, which hinders the galaxy from forming diffuse nebulae except through mergers with other galaxies.

In the dense nebulae where stars are produced, much of the hydrogen is in the molecular (H_2) form, so these nebulae are called molecular clouds. Observations indicate that the coldest clouds tend to form low-mass stars, observed first in the infrared inside the clouds, then in visible light at their surface when the clouds dissipate, while giant molecular clouds, which are generally warmer, produce stars of all masses. These giant molecular clouds have typical densities of 100 particles per cm^3, diameters of 100 light-years (9.5×10^{14} km), masses of up to 6 million solar masses ($M_\odot$), and an average interior temperature of 10 K. About half the total mass of the galactic ISM is found in molecular clouds and in the Milky Way there are an estimated 6,000 molecular clouds, each with more than 100,000 $M_\odot$. The nearest nebula to the Sun where massive stars are being formed is the Orion nebula, 1,300 ly (1.2×10^{16} km) away. However, lower mass star formation is occurring about 400–450 light years distant in the ρ Ophiuchi cloud complex.

A more compact site of star formation is the opaque clouds of dense gas and dust known as Bok globules; so named after the astronomer Bart Bok. These can form in association with collapsing molecular clouds or possibly independently. The Bok globules are typically up to a light year across and contain a few solar masses. They can be observed as dark clouds silhouetted against bright emission nebulae or background stars. Over half the known Bok globules have been found to contain newly forming stars.

Assembly of galaxy in early Universe.

Cloud Collapse

An interstellar cloud of gas will remain in hydrostatic equilibrium as long as the kinetic energy of the gas pressure is in balance with the potential energy of the internal gravitational force. Mathematically this is expressed using the virial theorem, which states that, to maintain equilibrium, the gravitational potential energy must equal twice the internal thermal energy. If a cloud is massive enough that the gas pressure is insufficient to support it, the cloud will undergo gravitational collapse. The mass above which a cloud will undergo such collapse is called the Jeans mass. The Jeans mass depends on the temperature and density of the cloud, but is typically thousands to tens of thousands of solar masses. This coincides with the typical mass of an open cluster of stars, which is the end product of a collapsing cloud.

Stellar cluster and star-forming region M 17.

In *triggered star formation*, one of several events might occur to compress a molecular cloud and initiate its gravitational collapse. Molecular clouds may collide with each other, or a nearby supernova explosion can be a trigger, sending shocked matter into the cloud at very high speeds. Alternatively, galactic collisions can trigger massive starbursts of star formation as the gas clouds in each galaxy are compressed and agitated by tidal forces. The latter mechanism may be responsible for the formation of globular clusters.

A supermassive black hole at the core of a galaxy may serve to regulate the rate of star formation in a galactic nucleus. A black hole that is accreting infalling matter can become active, emitting a strong wind through a collimated relativistic jet. This can limit further star formation. Massive black holes ejecting radio-frequency-emitting particles at near-light speed can also block the formation of new stars in aging galaxies. However, the radio emissions around the jets may also trigger star formation. Likewise, a weaker jet may trigger star formation when it collides with a cloud.

As it collapses, a molecular cloud breaks into smaller and smaller pieces in a hierarchical manner, until the fragments reach stellar mass. In each of these fragments, the collapsing gas radiates away the energy gained by the release of gravitational potential energy. As the density increases, the fragments become opaque and are thus less efficient at radiating away their energy. This raises the temperature of the cloud and inhibits further fragmentation. The fragments now condense into rotating spheres of gas that serve as stellar embryos.

Complicating this picture of a collapsing cloud are the effects of turbulence, macroscopic flows, rotation, magnetic fields and the cloud geometry. Both rotation and magnetic fields can hinder the collapse of a cloud. Turbulence is instrumental in causing fragmentation of the cloud, and on the smallest scales it promotes collapse.

Protostar

LH 95 stellar nursery in Large Magellanic Cloud.

A protostellar cloud will continue to collapse as long as the gravitational binding energy can be eliminated. This excess energy is primarily lost through radiation. However, the collapsing cloud will eventually become opaque to its own radiation, and the energy must be removed through some other means. The dust within the cloud becomes heated to temperatures of 60–100 K, and these particles radiate at wavelengths in the far infrared where the cloud is transparent. Thus the dust mediates the further collapse of the cloud.

During the collapse, the density of the cloud increases toward the center and thus the middle region becomes optically opaque first. This occurs when the density is about 10^{-13} g / cm^3. A core region, called the First Hydrostatic Core, forms where the collapse is essentially halted. It continues to increase in temperature as determined by the virial theorem. The gas falling toward this opaque region collides with it and creates shock waves that further heat the core.

Composite image showing young stars in and around molecular cloud Cepheus B.

When the core temperature reaches about 2000 K, the thermal energy dissociates the H_2 molecules. This is followed by the ionization of the hydrogen and helium atoms. These processes absorb the energy of the contraction, allowing it to continue on timescales comparable to the period of collapse at free fall velocities. After the density of infalling material has reached about 10^{-8} g / cm^3, that material is sufficiently transparent to allow energy radiated by the protostar to escape. The combination of convection within the protostar and radiation from its exterior allow the star to contract further. This continues until the gas is hot enough for the internal pressure to support the protostar against further gravitational collapse—a state called hydrostatic equilibrium. When this accretion phase is nearly complete, the resulting object is known as a protostar.

N11, part of a complex network of gas clouds and star clusters within our neighbouring galaxy, the Large Magellanic Cloud.

Accretion of material onto the protostar continues partially from the newly formed circumstellar disc. When the density and temperature are high enough, deuterium fusion begins, and the outward pressure of the resultant radiation slows (but does not stop) the collapse. Material comprising the cloud continues to "rain" onto the protostar. In this stage bipolar jets are produced called Herbig-Haro objects. This is probably the means by which excess angular momentum of the infalling material is expelled, allowing the star to continue to form.

When the surrounding gas and dust envelope disperses and accretion process stops, the star is considered a pre–main sequence star (PMS star). The energy source of these objects is gravitational contraction, as opposed to hydrogen burning in main sequence stars. The PMS star follows a Hayashi track on the Hertzsprung–Russell (H–R) diagram. The contraction will proceed until the Hayashi limit is reached, and thereafter contraction will continue on a Kelvin–Helmholtz timescale with the temperature remaining stable. Stars with less than 0.5 $M_{\odot}$ thereafter join the main sequence. For more massive PMS stars, at the end of the Hayashi track they will slowly collapse in near hydrostatic equilibrium, following the Henyey track.

Finally, hydrogen begins to fuse in the core of the star, and the rest of the enveloping material is cleared away. This ends the protostellar phase and begins the star's main sequence phase on the H–R diagram.

The stages of the process are well defined in stars with masses around 1 $M_{\odot}$ or less. In high mass stars, the length of the star formation process is comparable to the other timescales of their evolution, much shorter, and the process is not so well defined. The later evolution of stars are studied in stellar evolution.

Protostar

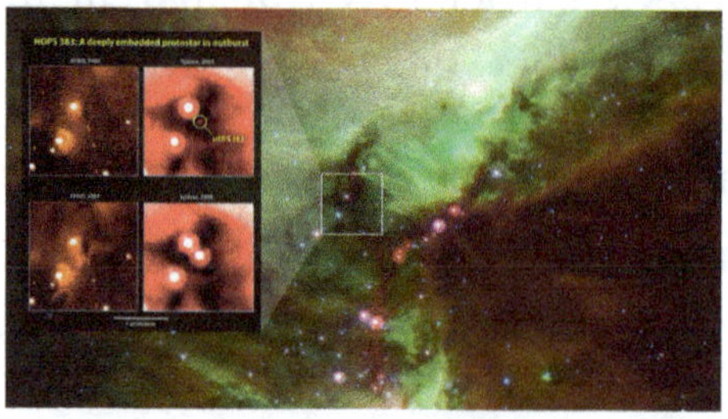

Protostar outburst - HOPS 383 (2015).

Observations

The Orion Nebula is an archetypical example of star formation, from the massive, young stars that are shaping the nebula to the pillars of dense gas that may be the homes of budding stars.

Key elements of star formation are only available by observing in wavelengths other than the optical. The protostellar stage of stellar existence is almost invariably hidden away deep inside dense clouds of gas and dust left over from the GMC. Often, these star-forming cocoons known as Bok globules, can be seen in silhouette against bright emission from surrounding gas. Early stages of a star's life can be seen in infrared light, which penetrates the dust more easily than visible light. Observations from the Wide-field Infrared Survey Explorer (WISE) have thus been especially important for unveiling numerous Galactic protostars and their parent star clusters. Examples of such embedded star clusters are FSR 1184, FSR 1190, Camargo 14, Camargo 74, Majaess 64, and Majaess 98.

Star-forming Region S106.

The structure of the molecular cloud and the effects of the protostar can be observed in near-IR extinction maps (where the number of stars are counted per unit area and compared to a nearby zero extinction area of sky), continuum dust emission and rotational transitions of CO and other molecules; these last two are observed in the millimeter and submillimeter range. The radiation from the protostar and early star has to be observed in infrared astronomy wavelengths, as the extinction caused by the rest of the cloud in which the star is forming is usually too big to allow us to observe it in the visual part of the spectrum. This presents considerable difficulties as the Earth's atmosphere is almost entirely opaque from 20μm to 850μm, with narrow windows at 200μm and 450μm. Even outside this range, atmospheric subtraction techniques must be used.

Young stars (purple) revealed by X-ray inside the NGC 2024 star-forming region.

X-ray observations have proven useful for studying young stars, since X-ray emission from these objects is 100–100,000 times stronger than X-ray emission from main-sequence stars. The earliest detections of X-rays from T Tauri stars were made by the Einstein X-ray Observatory. For low-mass stars X-rays are generated by the heating of the stellar corona through magnetic reconnection, while for high-mass O and early B-type stars X-rays are generated through supersonic shocks in the stellar winds. Photons in the soft X-ray energy range covered by the Chandra X-ray Observatory and XMM Newton may penetrate the interstellar medium with only moderate absorption due to gas, making the X-ray a useful wavelength for seeing the stellar populations within molecular clouds. X-ray emission as evidence of stellar youth makes this band particularly useful for performing censuses of stars in star-forming regions, given that not all young stars have infrared excesses. X-ray observations have provided near-complete censuses of all stellar-mass objects in the Orion Nebula Cluster and Taurus Molecular Cloud.

The formation of individual stars can only be directly observed in the Milky Way Galaxy, but in distant galaxies star formation has been detected through its unique spectral signature.

The first observed newborn star-forming clump, aged less than 10 million years old, was found in a galaxy about a corresponding light travel distance of 10.4 billion light years away, at an age when the universe was about 3.3 billion years old. The clump is about 3,000 light-years wide, and has a mass more than 1 billion times the mass of the sun, creating 32 stars each year with the mass of the sun, and produced about 40 percent of the stars in the clump's host galaxy.

Initial research indicates star-forming clumps start as giant, dense areas in turbulent gas-rich matter in young galaxies, live about 500 million years, and may migrate to the center of a galaxy, creating the central bulge of a galaxy.

On February 21, 2014, NASA announced a greatly upgraded database for tracking polycyclic aromatic hydrocarbons (PAHs) in the universe. According to scientists, more than 20% of the carbon in the universe may be associated with PAHs, possible starting materials for the formation of life. PAHs seem to have been formed shortly after the Big Bang, are widespread throughout the universe, and are associated with new stars and exoplanets.

Notable Pathfinder Objects

- MWC 349 was first discovered in 1978, and is estimated to be only 1,000 years old.
- VLA 1623 – The first exemplar Class 0 protostar, a type of embedded protostar that has yet to accrete the majority of its mass. Found in 1993, is possibly younger than 10,000 years.
- L1014 – An incredibly faint embedded object representative of a new class of sources that are only now being detected with the newest telescopes. Their status is still undetermined, they could be the youngest low-mass Class 0 protostars yet seen or even very low-mass evolved objects (like a brown dwarf or even an interstellar planet).
- IRS 8* – The youngest known main sequence star in the Galactic Center region, discovered in August 2006. It is estimated to be 3.5 million years old.

Low Mass and High Mass Star Formation

Star-forming region Westerhout 40 and the Serpens-Aquila Rift- cloud filaments containing new stars fill the region.

Stars of different masses are thought to form by slightly different mechanisms. The theory of low-mass star formation, which is well-supported by a plethora of observations, suggests that low-mass stars form by the gravitational collapse of rotating density enhancements within molecular clouds. As described above, the collapse of a rotating cloud of gas and dust leads to the formation of an accretion disk through which matter is channeled onto a central protostar. For stars with masses higher than about 8 $M_{\odot}$, however, the mechanism of star formation is not well understood.

Massive stars emit copious quantities of radiation which pushes against infalling material. In the past, it was thought that this radiation pressure might be substantial enough to halt accretion onto the massive protostar and prevent the formation of stars with masses more than a few tens of solar masses. Recent theoretical work has shown that the production of a jet and outflow clears a cavity through which much of the radiation from a massive protostar can escape without hindering accretion through the disk and onto the protostar. Present thinking is that massive stars may therefore be able to form by a mechanism similar to that by which low mass stars form.

There is mounting evidence that at least some massive protostars are indeed surrounded by accretion disks. Several other theories of massive star formation remain to be tested observationally. Of these, perhaps the most prominent is the theory of competitive accretion, which suggests that massive protostars are "seeded" by low-mass protostars which compete with other protostars to draw in matter from the entire parent molecular cloud, instead of simply from a small local region.

Another theory of massive star formation suggests that massive stars may form by the coalescence of two or more stars of lower mass.

Stellar Evolution

Mass (solar masses)	Time (years)	Spectral type
60	3 million	O3
30	11 million	O7
10	32 million	B4
3	370 million	A5
1.5	3 billion	F5
1	10 billion	G2 (Sun)
0.1	1000s billions	M7

Representative lifetimes of stars as a function of their masses.

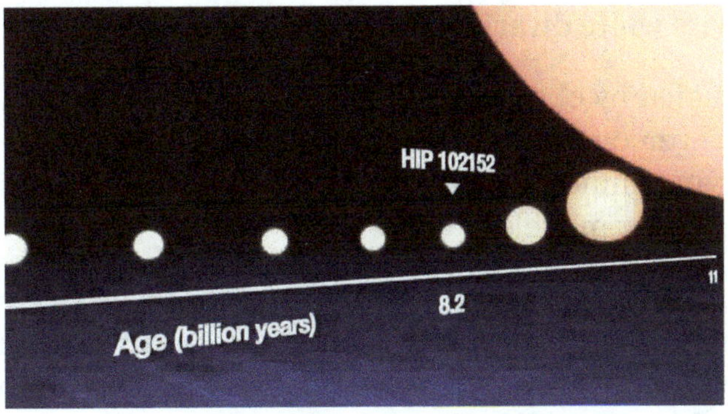

The life cycle of a Sun-like star.

Artist's depiction of the life cycle of a Sun-like star, starting as a main-sequence star at lower left then expanding through the subgiant and giant phases, until its outer envelope is expelled to form a planetary nebula at upper right.

Stellar evolution is the process by which a star changes over the course of time. Depending on the mass of the star, its lifetime can range from a few million years for the most massive to trillions of years for the least massive, which is considerably longer than the age of the universe. The table shows the lifetimes of stars as a function of their masses. All stars are born from collapsing clouds of gas and dust, often called nebulae or molecular clouds. Over the course of millions of years, these protostars settle down into a state of equilibrium, becoming what is known as a main-sequence star.

Nuclear fusion powers a star for most of its life. Initially the energy is generated by the fusion of hydrogen atoms at the core of the main-sequence star. Later, as the preponderance of atoms at the core becomes helium, stars like the Sun begin to fuse hydrogen along a spherical shell surrounding the core. This process causes the star to gradually grow in size, passing through the subgiant stage until it reaches the red giant phase. Stars with at least half the mass of the Sun can also begin to generate energy through the fusion of helium at their core, whereas more-massive stars can fuse heavier elements along a series of concentric shells. Once a star like the Sun has exhausted its nuclear fuel, its core collapses into a dense white dwarf and the outer layers are expelled as a planetary nebula. Stars with around ten or more times the mass of the Sun can explode in a supernova as their inert iron cores collapse into an extremely dense neutron star or black hole. Although the universe is not old enough for any of the smallest red dwarfs to have reached the end of their lives, stellar models suggest they will slowly become brighter and hotter before running out of hydrogen fuel and becoming low-mass white dwarfs.

Stellar evolution is not studied by observing the life of a single star, as most stellar changes occur too slowly to be detected, even over many centuries. Instead, astrophysicists come to understand how stars evolve by observing numerous stars at various points in their lifetime, and by simulating stellar structure using computer models.

In June 2015, astronomers reported evidence for Population III stars in the Cosmos Redshift 7 galaxy at $z = 6.60$. Such stars are likely to have existed in the very early universe (i.e., at high redshift), and may have started the production of chemical elements heavier than hydrogen that are needed for the later formation of planets and life as we know it.

Birth of a Star

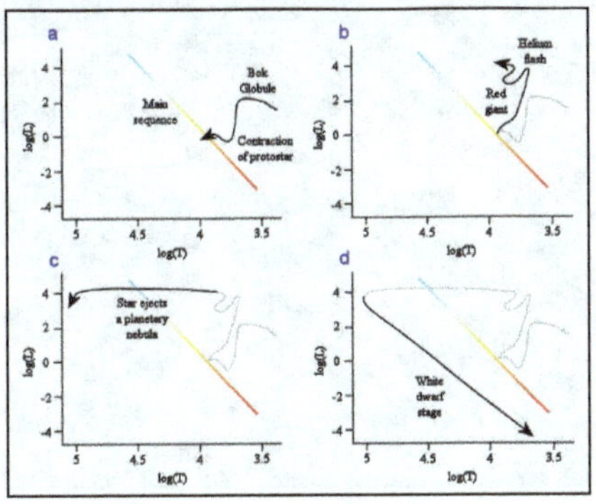

Schematic of stellar evolution.

Protostar

Stellar evolution starts with the gravitational collapse of a giant molecular cloud. Typical giant molecular clouds are roughly 100 light-years (9.5×10^{14} km) across and contain up to 6,000,000 solar masses (1.2×10^{37} kg). As it collapses, a giant molecular cloud breaks into smaller and smaller pieces. In each of these fragments, the collapsing gas releases gravitational potential energy as heat. As its temperature and pressure increase, a fragment condenses into a rotating sphere of superhot gas known as a protostar.

A protostar continues to grow by accretion of gas and dust from the molecular cloud, becoming a pre-main-sequence star as it reaches its final mass. Further development is determined by its mass. (Mass is compared to the mass of the Sun: 1.0 $M_\odot$ (2.0×10^{30} kg) means 1 solar mass.)

Protostars are encompassed in dust, and are thus more readily visible at infrared wavelengths. Observations from the Wide-field Infrared Survey Explorer (WISE) have been especially important for unveiling numerous Galactic protostars and their parent star clusters.

Brown Dwarfs and Sub-stellar Objects

Protostars with masses less than roughly 0.08 $M_\odot$ (1.6×10^{29} kg) never reach temperatures high enough for nuclear fusion of hydrogen to begin. These are known as brown dwarfs. The International Astronomical Union defines brown dwarfs as stars massive enough to fuse deuterium at some point in their lives (13 Jupiter masses (M_J), 2.5×10^{28} kg, or 0.0125 $M_\odot$). Objects smaller than 13 M_J are classified as sub-brown dwarfs (but if they orbit around another stellar object they are classified as planets). Both types, deuterium-burning and not, shine dimly and die away slowly, cooling gradually over hundreds of millions of years.

For a more-massive protostar, the core temperature will eventually reach 10 million kelvin, initiating the proton–proton chain reaction and allowing hydrogen to fuse, first to deuterium and then to helium. In stars of slightly over 1 $M_\odot$ (2.0×10^{30} kg), the carbon–nitrogen–oxygen fusion reaction (CNO cycle) contributes a large portion of the energy generation. The onset of nuclear fusion leads relatively quickly to a hydrostatic equilibrium in which energy released by the core exerts a "radiation pressure" balancing the weight of the star's matter, preventing further gravitational collapse. The star thus evolves rapidly to a stable state, beginning the main-sequence phase of its evolution.

A dense starfield in Sagittarius.

A new star will sit at a specific point on the main sequence of the Hertzsprung–Russell diagram, with the main-sequence spectral type depending upon the mass of the star. Small, relatively cold, low-mass red dwarfs fuse hydrogen slowly and will remain on the main sequence for hundreds of billions of years or longer, whereas massive, hot O-type stars will leave the main sequence after just a few million years. A mid-sized yellow dwarf star, like the Sun, will remain on the main sequence for about 10 billion years. The Sun is thought to be in the middle of its main sequence lifespan.

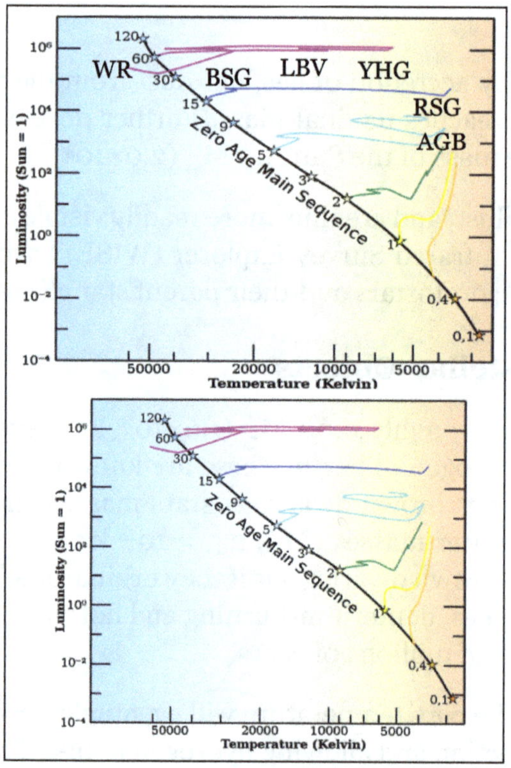

The evolutionary tracks of stars with different initial masses on the Hertzsprung–Russell diagram. The tracks start once the star has evolved to the main sequence and stop when fusion stops (for massive stars) and at the end of the red giant branch (for stars 1 $M_\odot$ and less).

A yellow track is shown for the Sun, which will become a red giant after its main-sequence phase ends before expanding further along the asymptotic giant branch, which will be the last phase in which the Sun undergoes fusion.

Mature Stars

Eventually the core exhausts its supply of hydrogen and the star begins to evolve off of the main sequence. Without the outward pressure generated by the fusion of hydrogen to counteract the force of gravity the core contracts until either electron degeneracy pressure becomes sufficient to oppose gravity or the core becomes hot enough (around 100 MK) for helium fusion to begin. Which of these happens first depends upon the star's mass.

Low-mass Stars

What happens after a low-mass star ceases to produce energy through fusion has not been directly observed; the universe is around 13.8 billion years old, which is less time (by several orders of magnitude, in some cases) than it takes for fusion to cease in such stars.

Recent astrophysical models suggest that red dwarfs of 0.1 $M_\odot$ may stay on the main sequence for some six to twelve trillion years, gradually increasing in both temperature and luminosity, and take several hundred billion more to collapse, slowly, into a white dwarf. Such stars will not become red giants as they are fully convective and will not develop a degenerate helium core with a shell burning hydrogen. Instead, hydrogen fusion will proceed until almost the whole star is helium.

Slightly more massive stars do expand into red giants, but their helium cores are not massive enough to reach the temperatures required for helium fusion so they never reach the tip of the red giant branch. When hydrogen shell burning finishes, these stars move directly off the red giant branch like a post AGB star, but at lower luminosity, to become a white dwarf. A star of about 0.5 $M_\odot$ will be able to reach temperatures high enough to fuse helium, and these "mid-sized" stars go on to further stages of evolution beyond the red giant branch.

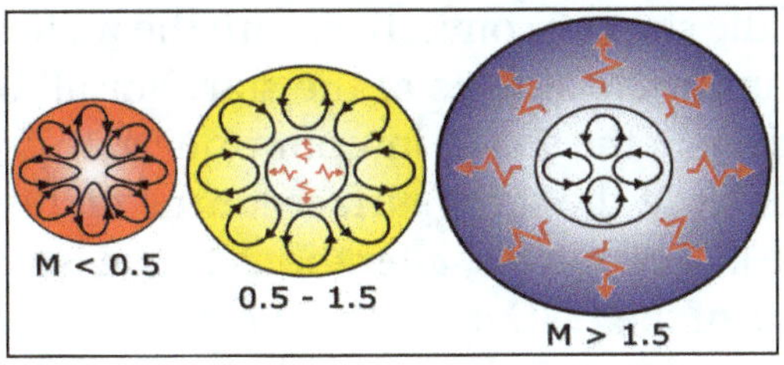

Internal structures of main-sequence stars, convection zones with arrowed cycles and radiative zones with red flashes. To the left a **low-mass** red dwarf, in the center a **mid-sized** yellow dwarf and at the right a **massive** blue-white main-sequence star.

Mid-sized Stars

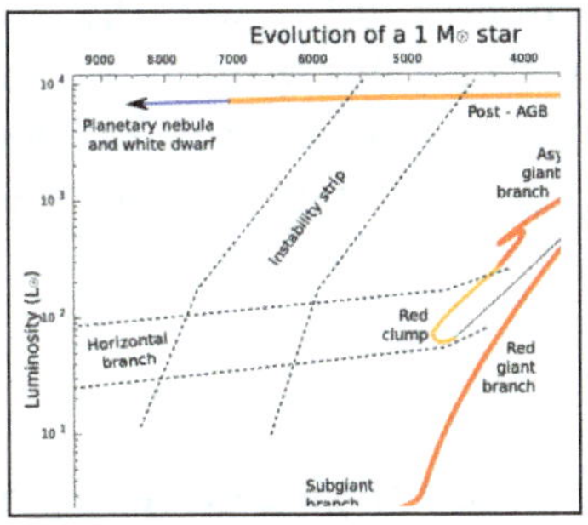

The evolutionary track of a solar mass, solar metallicity, star from main sequence to post-AGB.

Stars of roughly 0.5–10 $M_{\odot}$ become red giants, which are large non-main-sequence stars of stellar classification K or M. Red giants lie along the right edge of the Hertzsprung–Russell diagram due to their red color and large luminosity. Examples include Aldebaran in the constellation Taurus and Arcturus in the constellation of Boötes.

Mid-sized stars are red giants during two different phases of their post-main-sequence evolution: red-giant-branch stars, whose inert cores are made of helium, and asymptotic-giant-branch stars, whose inert cores are made of carbon. Asymptotic-giant-branch stars have helium-burning shells inside the hydrogen-burning shells, whereas red-giant-branch stars have hydrogen-burning shells only. Between these two phases, stars spend a period on the horizontal branch with a helium-fusing core. Many of these helium-fusing stars cluster towards the cool end of the horizontal branch as K-type giants and are referred to red clump giants.

Subgiant Phase

When a star exhausts the hydrogen in its core, it leaves the main sequence and begins to fuse hydrogen in a shell outside the core. The core increases in mass as the shell produces more helium. Depending on the mass of the helium core, this continues for several million to one or two billion years, with the star expanding and cooling at a similar or slightly lower luminosity to its main sequence state. Eventually either the core becomes degenerate, in stars around the mass of the sun, or the outer layers cool sufficiently to become opaque, in more massive stars. Either of these changes cause the hydrogen shell to increase in temperature and the luminosity of the star to increase, at which point the star expands onto the red giant branch.

Red-giant-branch Phase

The expanding outer layers of the star are convective, with the material being mixed by turbulence from near the fusing regions up to the surface of the star. For all but the lowest-mass stars, the fused material has remained deep in the stellar interior prior to this point, so the convecting envelope makes fusion products visible at the star's surface for the first time. At this stage of evolution, the results are subtle, with the largest effects, alterations to the isotopes of hydrogen and helium, being unobservable. The effects of the CNO cycle appear at the surface during the first dredge-up, with lower $^{12}C/^{13}C$ ratios and altered proportions of carbon and nitrogen. These are detectable with spectroscopy and have been measured for many evolved stars.

The helium core continues to grow on the red giant branch. It is no longer in thermal equilibrium, either degenerate or above the Schoenberg-Chandrasekhar limit, so it increases in temperature which causes the rate of fusion in the hydrogen shell to increase. The star increases in luminosity towards the tip of the red-giant branch. Red giant branch stars with a degenerate helium core all reach the tip with very similar core masses and very similar luminosities, although the more massive of the red giants become hot enough to ignite helium fusion before that point.

Horizontal Branch

If the core is largely supported by electron degeneracy pressure, helium fusion will ignite on a timescale of days in a helium flash. In more massive stars, the ignition of helium fusion occurs relatively slowly with no flash. The nuclear power released during the helium flash is very large, on the order of 10^8 times the luminosity of the Sun for a few days and 10^{11} times the luminosity of the Sun (roughly the luminosity of the Milky Way Galaxy) for a few seconds. However, the energy is absorbed by the stellar envelope and thus cannot be seen from outside the star. The energy released by helium fusion causes the core to expand, so that hydrogen fusion in the overlying layers slows and total energy generation decreases. The star contracts, although not all the way to the main sequence, and it migrates to the horizontal branch on the Hertzsprung–Russell diagram, gradually shrinking in radius and increasing its surface temperature.

Core helium flash stars evolve to the red end of the horizontal branch but do not migrate to higher temperatures before they gain a degenerate carbon-oxygen core and start helium shell burning. These stars are often observed as a red clump of stars in the colour-magnitude diagram of a cluster, hotter and less luminous than the red giants. Higher-mass stars with larger helium cores move along the horizontal branch to higher temperatures, some becoming unstable pulsating stars in the yellow instability strip (RR Lyrae variables), whereas some become even hotter and can form a blue tail or blue hook to the horizontal branch. The exact morphology of the horizontal branch depends on parameters such as metallicity, age, and helium content, but the exact details are still being modelled.

Asymptotic-giant-branch Phase

After a star has consumed the helium at the core, hydrogen and helium fusion continues in shells around a hot core of carbon and oxygen. The star follows the asymptotic giant branch on the Hertzsprung–Russell diagram, paralleling the original red giant evolution, but with even faster energy generation (which lasts for a shorter time). Although helium is being burnt in a shell, the majority of the energy is produced by hydrogen burning in a shell further from the core of the star. Helium from these hydrogen burning shells drops towards the center of the star and periodically the energy output from the helium shell increases dramatically. This is known as a thermal pulse and they occur towards the end of the asymptotic-giant-branch phase, sometimes even into the post-asymptotic-giant-branch phase. Depending on mass and composition, there may be several to hundreds of thermal pulses.

There is a phase on the ascent of the asymptotic-giant-branch where a deep convective zone forms and can bring carbon from the core to the surface. This is known as the second dredge up, and in some stars there may even be a third dredge up. In this way a carbon star is formed, very cool and strongly reddened stars showing strong carbon lines in their spectra. A process known as hot bottom burning may convert carbon into oxygen and nitrogen before it can be dredged to the surface, and the interaction between these processes determines the observed luminosities and spectra of carbon stars in particular clusters.

Another well known class of asymptotic-giant-branch stars are the Mira variables, which pulsate with well-defined periods of tens to hundreds of days and large amplitudes up to about 10 magnitudes (in the visual, total luminosity changes by a much smaller amount). In more-massive stars the stars become more luminous and the pulsation period is longer, leading to enhanced mass loss, and the stars become heavily obscured at visual wavelengths. These stars can be observed as OH/IR stars, pulsating in the infra-red and showing OH maser activity. These stars are clearly oxygen rich, in contrast to the carbon stars, but both must be produced by dredge ups.

Post-AGB

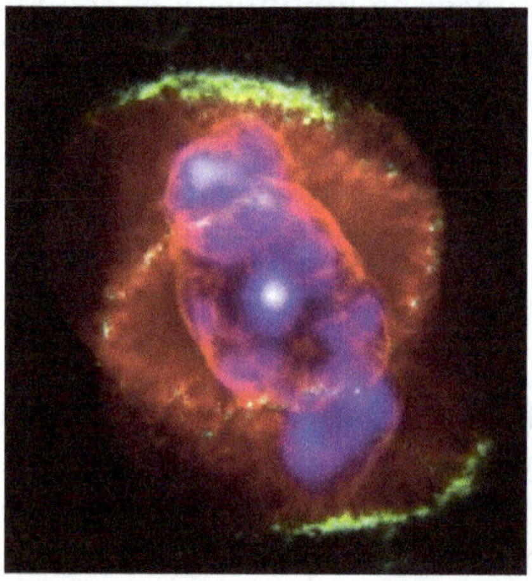

The Cat's Eye Nebula, a planetary nebula formed by the death of a star with about the same mass as the Sun.

These mid-range stars ultimately reach the tip of the asymptotic-giant-branch and run out of fuel for shell burning. They are not sufficiently massive to start full-scale carbon fusion, so they contract again, going through a period of post-asymptotic-giant-branch superwind to produce a planetary nebula with an extremely hot central star. The central star then cools to a white dwarf. The expelled gas is relatively rich in heavy elements created within the star and may be particularly oxygen or carbon enriched, depending on the type of the star. The gas builds up in an expanding shell called a circumstellar envelope and cools as it moves away from the star, allowing dust particles and molecules to form. With the high infrared energy input from the central star, ideal conditions are formed in these circumstellar envelopes for maser excitation.

It is possible for thermal pulses to be produced once post-asymptotic-giant-branch evolution has begun, producing a variety of unusual and poorly understood stars known as born-again asymptotic-giant-branch stars. These may result in extreme horizontal-branch stars (subdwarf B stars), hydrogen deficient post-asymptotic-giant-branch stars, variable planetary nebula central stars, and R Coronae Borealis variables.

Massive Stars

The Crab Nebula, the shattered remnants of a star which exploded as a supernova, the light of which reached Earth in 1054 AD.

In massive stars, the core is already large enough at the onset of the hydrogen burning shell that helium ignition will occur before electron degeneracy pressure has a chance to become prevalent. Thus, when these stars expand and cool, they do not brighten as much as lower-mass stars; however, they were much brighter than lower-mass stars to begin with, and are thus still brighter than the red giants formed from less-massive stars. These stars are unlikely to survive as red supergiants; instead they will destroy themselves as type II supernovas.

Extremely massive stars (more than approximately 40 $M_\odot$), which are very luminous and thus have very rapid stellar winds, lose mass so rapidly due to radiation pressure that they tend to strip off their own envelopes before they can expand to become red supergiants, and thus retain extremely high surface temperatures (and blue-white color) from their main-sequence time onwards. The largest stars of the current generation are about 100-150 $M_\odot$ because the outer layers would be expelled by the extreme radiation. Although lower-mass stars normally do not burn off their outer layers so rapidly, they can likewise avoid becoming red giants or red supergiants if they are in binary systems close enough so that the companion star strips off the envelope as it expands, or if they rotate rapidly enough so that convection extends all the way from the core to the surface, resulting in the absence of a separate core and envelope due to thorough mixing.

The core grows hotter and denser as it gains material from fusion of hydrogen at the base of the envelope. In all massive stars, electron degeneracy pressure is insufficient to halt collapse by itself, so as each major element is consumed in the center, progressively heavier elements ignite, temporarily halting collapse. If the core of the star is not too massive (less than approximately 1.4 $M_\odot$, taking into account mass loss that has occurred by this time), it may then form a white dwarf (possibly surrounded by a planetary nebula) as described above for less-massive stars, with the difference that the white dwarf is composed chiefly of oxygen, neon, and magnesium.

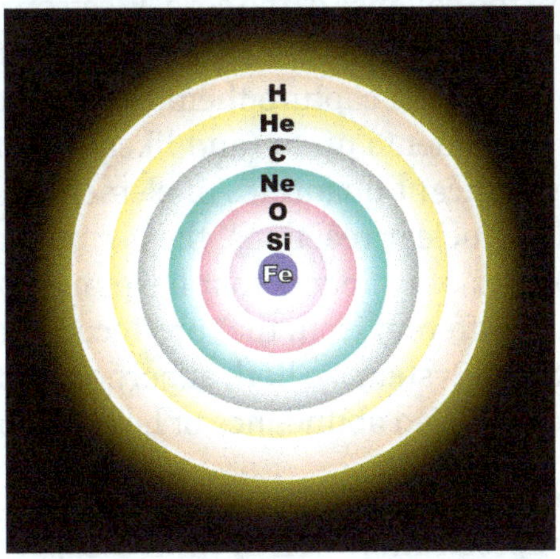

The onion-like layers of a massive, evolved star just before core collapse. (Not to scale.)

Above a certain mass (estimated at approximately 2.5 $M_\odot$ and whose star's progenitor was around 10 $M_\odot$), the core will reach the temperature (approximately 1.1 gigakelvins) at which neon partially breaks down to form oxygen and helium, the latter of which immediately fuses with some of the remaining neon to form magnesium; then oxygen fuses to form sulfur, silicon, and smaller amounts of other elements. Finally, the temperature gets high enough that any nucleus can be partially broken down, most commonly releasing an alpha particle (helium nucleus) which immediately fuses with another nucleus, so that several nuclei are effectively rearranged into a smaller number of heavier nuclei, with net release of energy because the addition of fragments to nuclei exceeds the energy required to break them off the parent nuclei.

A star with a core mass too great to form a white dwarf but insufficient to achieve sustained conversion of neon to oxygen and magnesium, will undergo core collapse (due to electron capture) before achieving fusion of the heavier elements. Both heating and cooling caused by electron capture onto minor constituent elements (such as aluminum and sodium) prior to collapse may have a significant impact on total energy generation within the star shortly before collapse. This may produce a noticeable effect on the abundance of elements and isotopes ejected in the subsequent supernova.

Supernova

Once the nucleosynthesis process arrives at iron-56, the continuation of this process consumes energy (the addition of fragments to nuclei releases less energy than required to break them off the parent nuclei). If the mass of the core exceeds the Chandrasekhar limit, electron degeneracy pressure will be unable to support its weight against the force of gravity, and the core will undergo sudden, catastrophic collapse to form a neutron star or (in the case of cores that exceed the Tolman-Oppenheimer-Volkoff limit), a black hole. Through a process that is not completely understood, some of the gravitational potential energy released by this core collapse is converted into a Type Ib, Type Ic, or Type II supernova. It is known that the core collapse produces a massive surge of neutrinos, as observed with supernova SN 1987A. The extremely energetic neutrinos fragment some nuclei; some of their energy is consumed in releasing nucleons, including neutrons, and some of their energy is transformed into heat and kinetic energy, thus augmenting the shock wave started by rebound of some of the infalling material from the collapse of the core. Electron capture in very dense parts of the infalling matter may produce additional neutrons. Because some of the rebounding matter is bombarded by the neutrons, some of its nuclei capture them, creating a spectrum of heavier-than-iron material including the radioactive elements up to (and likely beyond) uranium. Although non-exploding red giants can produce significant quantities of elements heavier than iron using neutrons released in side reactions of earlier nuclear reactions, the abundance of elements heavier than iron (and in particular, of certain isotopes of elements that have multiple stable or long-lived isotopes) produced in such reactions is quite different from that produced in a supernova. Neither abundance alone matches that found in the Solar System, so both supernovae and ejection of elements from red giants are required to explain the observed abundance of heavy elements and isotopes thereof.

The energy transferred from collapse of the core to rebounding material not only generates heavy elements, but provides for their acceleration well beyond escape velocity, thus causing a Type Ib, Type Ic, or Type II supernova. Note that current understanding of this energy transfer is still not satisfactory; although current computer models of Type Ib, Type Ic, and Type II supernovae account for part of the energy transfer, they are not able to account for enough energy transfer to produce the observed ejection of material.

Some evidence gained from analysis of the mass and orbital parameters of binary neutron stars (which require two such supernovae) hints that the collapse of an oxygen-neon-magnesium core may produce a supernova that differs observably (in ways other than size) from a supernova produced by the collapse of an iron core.

The most-massive stars that exist today may be completely destroyed by a supernova with an energy greatly exceeding its gravitational binding energy. This rare event, caused by pair-instability, leaves behind no black hole remnant. In the past history of the universe, some stars were even larger than the largest that exists today, and they would immediately collapse into a black hole at the end of their lives, due to photodisintegration.

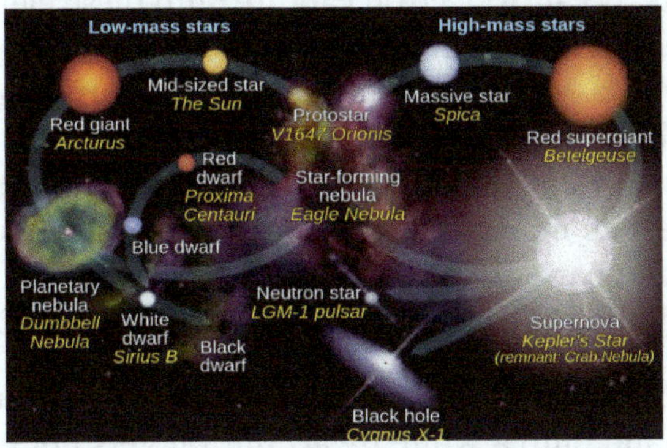

Stellar evolution of low-mass (left cycle) and high-mass (right cycle) stars, with examples in italics.

Stellar Remnants

After a star has burned out its fuel supply, its remnants can take one of three forms, depending on the mass during its lifetime.

White and Black Dwarfs

For a star of 1 $M_{\odot}$, the resulting white dwarf is of about 0.6 $M_{\odot}$, compressed into approximately the volume of the Earth. White dwarfs are stable because the inward pull of gravity is balanced by the degeneracy pressure of the star's electrons, a consequence of the Pauli exclusion principle. Electron degeneracy pressure provides a rather soft limit against further compression; therefore, for a given chemical composition, white dwarfs of higher mass have a smaller volume. With no fuel left to burn, the star radiates its remaining heat into space for billions of years.

A white dwarf is very hot when it first forms, more than 100,000 K at the surface and even hotter in its interior. It is so hot that a lot of its energy is lost in the form of neutrinos for the first 10 million years of its existence, but will have lost most of its energy after a billion years.

The chemical composition of the white dwarf depends upon its mass. A star of a few solar masses will ignite carbon fusion to form magnesium, neon, and smaller amounts of other elements, resulting in a white dwarf composed chiefly of oxygen, neon, and magnesium, provided that it can lose enough mass to get below the Chandrasekhar limit, and provided that the ignition of carbon is not so violent as to blow the star apart in a supernova. A star of mass on the order of magnitude of the Sun will be unable to ignite carbon fusion, and will produce a white dwarf composed chiefly of carbon and oxygen, and of mass too low to collapse unless matter is added to it later. A star of less than about half the mass of the Sun will be unable to ignite helium fusion (as noted earlier), and will produce a white dwarf composed chiefly of helium.

In the end, all that remains is a cold dark mass sometimes called a black dwarf. However, the universe is not old enough for any black dwarfs to exist yet.

If the white dwarf's mass increases above the Chandrasekhar limit, which is 1.4 $M_\odot$ for a white dwarf composed chiefly of carbon, oxygen, neon, and/or magnesium, then electron degeneracy pressure fails due to electron capture and the star collapses. Depending upon the chemical composition and pre-collapse temperature in the center, this will lead either to collapse into a neutron star or runaway ignition of carbon and oxygen. Heavier elements favor continued core collapse, because they require a higher temperature to ignite, because electron capture onto these elements and their fusion products is easier; higher core temperatures favor runaway nuclear reaction, which halts core collapse and leads to a Type Ia supernova. These supernovae may be many times brighter than the Type II supernova marking the death of a massive star, even though the latter has the greater total energy release. This instability to collapse means that no white dwarf more massive than approximately 1.4 $M_\odot$ can exist (with a possible minor exception for very rapidly spinning white dwarfs, whose centrifugal force due to rotation partially counteracts the weight of their matter). Mass transfer in a binary system may cause an initially stable white dwarf to surpass the Chandrasekhar limit.

If a white dwarf forms a close binary system with another star, hydrogen from the larger companion may accrete around and onto a white dwarf until it gets hot enough to fuse in a runaway reaction at its surface, although the white dwarf remains below the Chandrasekhar limit. Such an explosion is termed a nova.

Neutron Stars

Bubble-like shock wave still expanding from a supernova explosion 15,000 years ago.

Ordinarily, atoms are mostly electron clouds by volume, with very compact nuclei at the center (proportionally, if atoms were the size of a football stadium, their nuclei would be the size of dust mites). When a stellar core collapses, the pressure causes electrons and protons to fuse by electron capture. Without electrons, which keep nuclei apart, the neutrons collapse into a dense ball (in some ways like a giant atomic nucleus), with a thin overlying layer of degenerate matter (chiefly iron unless matter of different composition is added later). The neutrons resist further compression by the Pauli Exclusion Principle, in a way analogous to electron degeneracy pressure, but stronger.

These stars, known as neutron stars, are extremely small—on the order of radius 10 km, no bigger than the size of a large city—and are phenomenally dense. Their period of rotation shortens dramatically as the stars shrink (due to conservation of angular momentum); observed rotational periods of neutron stars range from about 1.5 milliseconds (over 600 revolutions per second) to several seconds. When these rapidly rotating stars' magnetic poles are aligned with the Earth, we detect a pulse of radiation each revolution. Such neutron stars are called pulsars, and were the first neutron stars to be discovered. Though electromagnetic radiation detected from pulsars is most often in the form of radio waves, pulsars have also been detected at visible, X-ray, and gamma ray wavelengths.

Black Holes

If the mass of the stellar remnant is high enough, the neutron degeneracy pressure will be insufficient to prevent collapse below the Schwarzschild radius. The stellar remnant thus becomes a black hole. The mass at which this occurs is not known with certainty, but is currently estimated at between 2 and 3 $M_\odot$.

Black holes are predicted by the theory of general relativity. According to classical general relativity, no matter or information can flow from the interior of a black hole to an outside observer, although quantum effects may allow deviations from this strict rule. The existence of black holes in the universe is well supported, both theoretically and by astronomical observation.

Because the core-collapse supernova mechanism itself is imperfectly understood, it is still not known whether it is possible for a star to collapse directly to a black hole without producing a visible supernova, or whether some supernovae initially form unstable neutron stars which then collapse into black holes; the exact relation between the initial mass of the star and the final remnant is also not completely certain. Resolution of these uncertainties requires the analysis of more supernovae and supernova remnants.

Models

A stellar evolutionary model is a mathematical model that can be used to compute the evolutionary phases of a star from its formation until it becomes a remnant. The mass and chemical composition of the star are used as the inputs, and the luminosity and surface temperature are the only constraints. The model formulae are based upon the physical understanding of the star, usually under the assumption of hydrostatic equilibrium. Extensive computer calculations are then run to determine the changing state of the star over time, yielding a table of data that can be used to determine the evolutionary track of the star across the Hertzsprung–Russell diagram, along with other evolving properties. Accurate models can be used to estimate the current age of a star by comparing its physical properties with those of stars along a matching evolutionary track.

Supernova

SN 1994D (bright spot on the lower left), a type Ia supernova in the NGC 4526 galaxy.

A supernova is an astronomical event that occurs during the last stellar evolutionary stages of a massive star's life, whose dramatic and catastrophic destruction is marked by one final titanic explosion. For a short time, this causes the sudden appearance of a 'new' bright star, before slowly fading from sight over several weeks or months.

Only three Milky Way naked-eye supernova events have been observed during the last thousand years, though many have been telescopically seen in other galaxies. The most recent directly observed supernova in the Milky Way was Kepler's Star in 1604 (SN 1604), but remnants of two more recent supernovae have been found retrospectively. Statistical observations of supernovae in other galaxies suggest they should occur on average about three times every century in the Milky Way, and that any galactic supernova would almost certainly be observable in modern astronomical equipment.

Supernovae are more energetic than novae. In Latin, *Nova* means "new", referring astronomically to what appears to be a temporary new bright star. Adding the prefix "super-" distinguishes supernovae from ordinary novae, which are far less luminous. The word *supernova* was coined by Walter Baade and Fritz Zwicky in 1931.

During maximum brightness, the total equivalent radiant energies produced by supernovae may briefly outshine an entire output of a typical galaxy and emit energies equal to that created over the lifetime of any solar-like star. Such extreme catastrophes may also expel much, if not all, of its stellar material away from the star, at velocities up to 30,000 km/s or 10% of the speed of light. This drives an expanding and fast-moving shock wave into the surrounding interstellar medium, and in turn, sweeping up an expanding shell of gas and dust, which is observed as a supernova remnant. Supernovae create, fuse and eject the bulk of the chemical elements produced by nucleosynthesis. Supernovae play a significant role in enriching the interstellar medium with the heavier atomic mass chemical elements. Furthermore, the expanding shock waves from supernova explosions can trigger the formation of new stars. Supernova remnants are expected to accelerate a large fraction of galactic primary cosmic rays, but direct evidence for cos-mic ray production was found only in a few of them so far. They are also potentially strong galactic sources of gravitational waves.

Theoretical studies of many supernovae indicate that most are triggered by one of two basic mechanisms: the sudden re-ignition of nuclear fusion in a degenerate star or the sudden gravitational collapse of a massive star's core. In the first instance, a degenerate white dwarf may accumulate sufficient material from a binary companion, either through accretion or via a merger, to raise its core temperature enough to trigger runaway nuclear fusion, completely disrupting the star. In the second case, the core of a massive star may undergo sudden gravitational collapse, releasing gravitational potential energy as a supernova. While some observed supernovae are more complex than these two simplified theories, the astrophysical collapse mechanics have been established and accepted by most astronomers for some time.

Due to the wide range of astrophysical consequences of these events, astronomers now deem supernovae research, across the fields of stellar and galactic evolution, as an especially important area for investigation.

Etymology

In Latin, *Nova* means "new", referring astronomically to what appears to be a temporary new bright star. Adding the prefix "super-" distinguishes supernovae from ordinary novae, which are far less luminous. The word *supernova* was coined by Walter Baade and Fritz Zwicky in 1931.

Observation History

The Crab Nebula is a pulsar wind nebula associated with the 1054 supernova.

The highlighted passages refer to the Chinese observation of SN 1054

The earliest recorded supernova, SN 185, was viewed by Chinese astronomers in 185 AD, with the brightest recorded supernova being SN 1006, which occurred in 1006 AD and was described in detail by Chinese and Islamic astronomers. The widely observed supernova SN 1054 produced the Crab Nebula. Supernovae SN 1572 and SN 1604, the latest to be observed with the naked eye in the Milky Way galaxy, had notable effects on the development of astronomy in Europe because they were used to argue against the Aristotelian idea that the universe beyond the Moon and planets was static and unchanging. Johannes Kepler began observing SN 1604 at its peak on October 17, 1604, and continued to make estimates of its brightness until it faded from naked eye view a year later. It was the second supernova to be observed in a generation (after SN 1572 seen by Tycho Brahe in Cassiopeia).

Before the development of the telescope, there have only been five supernovae seen in the last millennium. In the perspective of how long a star's lifetime is, its death is very brief. In fact, a star's death may only last a few months. Due to this, a typical human will only experience this rarity, on average, once in their lifetime. This is a microscopic fraction in comparison to the 100 billion stars that compose a galaxy. However, since the use of modern equipment, particularly in this millennium, professional and amateur astronomers have been finding several hundreds of supernovae each year.

The field of supernova discovery has extended to other galaxies, starting with SN 1885A in the Andromeda galaxy. American astronomers Rudolph Minkowski and Fritz Zwicky developed the modern supernova classification scheme beginning in 1941. In the 1960s, astronomers found that the maximum intensities of supernova explosions could be used as standard candles, hence indicators of astronomical distances. Some of the most distant supernovae recently observed appeared dimmer than expected. This supports the view that the expansion of the universe is accelerating. Techniques were developed for reconstructing supernova explosions that have no written records of being observed. The date of the Cassiopeia A supernova event was determined from light echoes off nebulae, while the age of supernova remnant RX J0852.0-4622 was estimated from temperature measurements and the gamma ray emissions from the decay of titanium-44.

The brightest observed supernova, ASASSN-15lh, was detected in June 2015. With a brightness of 570 billion Suns, ASASSN-15lh's peak luminosity was twice that of the previous record holder.

Discovery

Early work on what was originally believed to be simply a new category of novae was performed during the 1930s by Walter Baade and Fritz Zwicky at Mount Wilson Observatory. The name *super-novae* was first used during 1931 lectures held at Caltech by Baade and Zwicky, then used publicly in 1933 at a meeting of the American Physical Society. By 1938, the hyphen had been lost and the modern name was in use. Because supernovae are relatively rare events within a galaxy, occurring about three times a century in the Milky Way, obtaining a good sample of supernovae to study requires regular monitoring of many galaxies.

Supernovae in other galaxies cannot be predicted with any meaningful accuracy. Normally, when they are discovered, they are already in progress. Most scientific interest in supernovae—as standard candles for measuring distance, for example—require an observation of their peak luminosity. It is therefore important to discover them well before they reach their maximum. Amateur astronomers, who greatly outnumber professional astronomers, have played an important role in finding supernovae, typically by looking at some of the closer galaxies through an optical telescope and comparing them to earlier photographs.

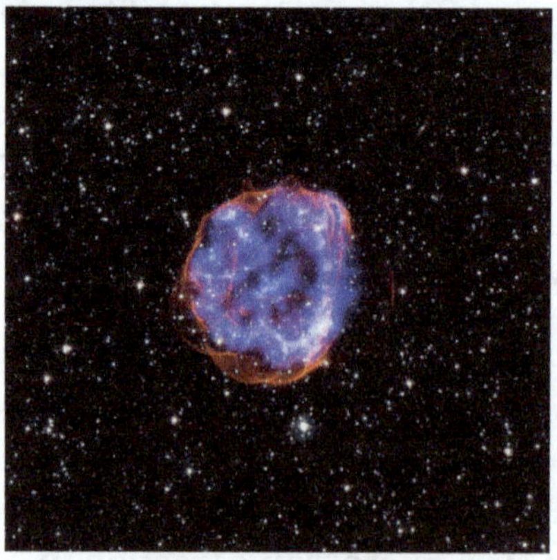

A supernova remnant.

Toward the end of the 20th century astronomers increasingly turned to computer-controlled telescopes and CCDs for hunting supernovae. While such systems are popular with amateurs, there are also professional installations such as the Katzman Automatic Imaging Telescope. Recently the Supernova Early Warning System (SNEWS) project has begun using a network of neutrino detectors to give early warning of a supernova in the Milky Way galaxy. Neutrinos are particles that are produced in great quantities by a supernova explosion, and they are not significantly absorbed by the interstellar gas and dust of the galactic disk.

"A star set to explode", the SBW1 nebula surrounds a massive blue supergiant in the Carina Nebula.

Supernova searches fall into two classes: those focused on relatively nearby events and those looking for explosions farther away. Because of the expansion of the universe, the distance to a remote object with a known emission spectrum can be estimated by measuring its Doppler shift (or redshift); on average, more distant objects recede with greater velocity than those nearby, and so have a higher redshift. Thus the search is split between high redshift and low redshift, with the boundary falling around a redshift range of $z = 0.1–0.3$—where z is a dimensionless measure of the spectrum's frequency shift.

High redshift searches for supernovae usually involve the observation of supernova light curves. These are useful for standard or calibrated candles to generate Hubble diagrams and make cosmological predictions. Supernova spectroscopy, used to study the physics and environments of supernovae, is more practical at low than at high redshift. Low redshift observations also anchor the low-distance end of the Hubble curve, which is a plot of distance versus redshift for visible galaxies.

Naming Convention

Multiwavelength X-ray, infrared, and optical compilation image of Kepler's supernova remnant, SN 1604.

Supernova discoveries are reported to the International Astronomical Union's Central Bureau for Astronomical Telegrams, which sends out a circular with the name it assigns to that supernova. The name is the marker *SN* followed by the year of discovery, suffixed with a one or two-letter designation. The first 26 supernovae of the year are designated with a capital letter from *A* to *Z*. Afterward pairs of lower-case letters are used: *aa*, *ab*, and so on. Hence, for example, *SN 2003C* designates the third supernova reported in the year 2003. The last supernova of 2005 was SN 2005nc, indicating that it was the 367th supernova found in 2005. Since 2000, professional and amateur astronomers have been finding several hundreds of supernovae each year (572 in 2007, 261 in 2008, 390 in 2009; 231 in 2013).

Historical supernovae are known simply by the year they occurred: SN 185, SN 1006, SN 1054, SN 1572 (called *Tycho's Nova*) and SN 1604 (*Kepler's Star*). Since 1885 the additional letter notation has been used, even if there was only one supernova discovered that year (e.g. SN 1885A, SN 1907A, etc.) — this last happened with SN 1947A. *SN*, for SuperNova, is a standard prefix. Until 1987, two-letter designations were rarely needed; since 1988, however, they have been needed every year.

Classification

As part of the attempt to understand supernovae, astronomers have classified them according to their light curves and the absorption lines of different chemical elements that appear in their spectra. The first element for division is the presence or absence of a line caused by hydrogen. If a supernova's spectrum contains lines of hydrogen (known as the Balmer series in the visual portion of the spectrum) it is classified *Type II*; otherwise it is *Type I*. In each of these two types there are subdivisions according to the presence of lines from other elements or the shape of the light curve (a graph of the supernova's apparent magnitude as a function of time).

Artist's impression of supernova 1993J.

<table>
<tr><th colspan="5">Supernova taxonomy</th></tr>
<tr><td rowspan="3">Type I
No hydrogen</td><td colspan="3">Type Ia
Presents a singly ionized silicon (Si II) line at 615.0 nm (nanometers), near peak light</td><td>Thermal runaway</td></tr>
<tr><td rowspan="2">Type Ib/c
Weak or no silicon absorption feature</td><td colspan="2">Type Ib
Shows a non-ionized helium (He I) line at 587.6 nm</td><td rowspan="6">Core collapse</td></tr>
<tr><td colspan="2">Type Ic
Weak or no helium</td></tr>
<tr><td rowspan="4">Type II
Shows hydrogen</td><td rowspan="3">Type II-P/L/N
Type II spectrum throughout</td><td rowspan="2">Type II-P/L
No narrow lines</td><td>Type II-P
Reaches a "plateau" in its light curve</td></tr>
<tr><td>Type II-L
Displays a "linear" decrease in its light curve (linear in magnitude versus time).</td></tr>
<tr><td colspan="2">Type IIn
Some narrow lines</td></tr>
<tr><td colspan="3">Type IIb
Spectrum changes to become like Type Ib</td></tr>
</table>

Type I

Type I supernovae are subdivided on the basis of their spectra, with type Ia showing a strong ionised silicon absorption line. Type I supernovae without this strong line are classified as Type Ib and Ic, with Type Ib showing strong neutral helium lines and Type Ic lacking them. The light curves are all similar, although Type Ia are generally brighter at peak luminosity, but the light curve is not important for classification of Type I supernovae.

A small number of Type Ia supernovae exhibit unusual features such as non-standard luminosity or broadened light curves, and these are typically classified by referring to the earliest example showing similar features. For example, the sub-luminous SN 2008ha is often referred to as SN 2002cx-like or class Ia-2002cx.

Type II

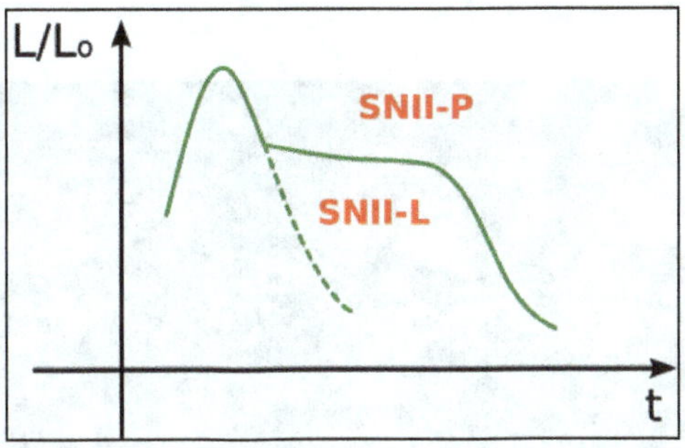

Light curves are used to classify type II-P and type II-L supernovae.

The supernovae of Type II can also be sub-divided based on their spectra. While most Type II supernovae show very broad emission lines which indicate expansion velocities of many thousands of kilometres per second, some, such as SN 2005gl, have relatively narrow features in their spectra. These are called Type IIn, where the 'n' stands for 'narrow'.

A few supernovae, such as SN 1987K and SN 1993J, appear to change types: they show lines of hydrogen at early times, but, over a period of weeks to months, become dominated by lines of helium. The term "Type IIb" is used to describe the combination of features normally associated with Types II and Ib.

Type II supernovae with normal spectra dominated by broad hydrogen lines that remain for the life of the decline are classified on the basis of their light curves. The most common type shows a distinctive "plateau" in the light curve shortly after peak brightness where the visual luminosity stays relatively constant for several months before the decline resumes. These are called type II-P referring to the plateau. Less common are type II-L supernovae that lack a distinct plateau. The "L" signifies "linear" although the light curve is not actually a straight line.

Supernovae that do not fit into the normal classifications are designated peculiar, or 'pec'.

Types III, IV, and V

Fritz Zwicky defined additional supernovae types, although based on a very few examples that didn't cleanly fit the parameters for a Type I or Type II supernova. SN 1961i in NGC 4303 was the prototype and only member of the Type III supernova class, noted for its broad light curve maximum and broad hydrogen Balmer lines that were slow to develop in the spectrum. SN 1961f in NGC 3003 was the prototype and only member of the Type IV class, with a light curve similar to a Type II-P supernova, with hydrogen absorption lines but weak hydrogen emission lines. The Type V class was coined for SN 1961V in NGC 1058, an unusual faint supernova or supernova imposter with a slow rise to brightness, a maximum lasting many months, and an unusual emission spectrum. The similarity of SN 1961V to the Eta Carinae Great Outburst was noted. Supernovae in M101 (1909) and M83 (1923 and 1957) were also suggested as possible type IV or type V supernovae.

These types would now all be treated as peculiar Type II supernovae, of which many more examples have been discovered, although it is still debated whether SN 1961V was a true supernova following an LBV outburst or an imposter.

Current Models

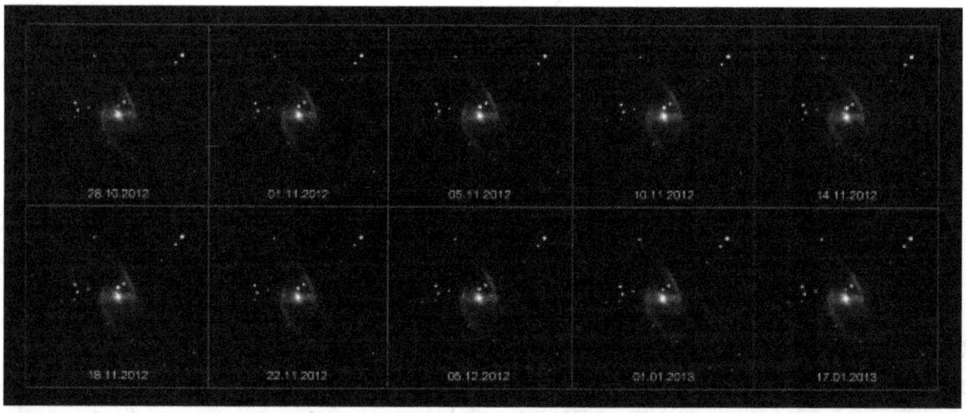

Sequence shows the rapid brightening and slower fading of a supernova explosion in the galaxy NGC 1365.

The type codes, described above given to supernovae, are *taxonomic* in nature: the type number describes the light observed from the supernova, not necessarily its cause. For example, type Ia supernovae are produced by runaway fusion ignited on degenerate white dwarf progenitors while the spectrally similar type Ib/c are produced from massive Wolf-Rayet progenitors by core collapse. The following summarizes what is currently believe to be the most plausible explanations for supernovae.

Thermal Runaway

A white dwarf star may accumulate sufficient material from a stellar companion to raise its core temperature enough to ignite carbon fusion, at which point it undergoes runaway nuclear fusion, completely disrupting it. There are three avenues by which this detonation is theorized to happen: stable accretion of material from a companion, the collision of two white dwarfs, or accretion that causes ignition in a shell that then ignites. The dominant mechanism by which Type Ia supernovae are produced remains unclear. Despite this uncertainty in how Type Ia supernovae are produced, Type Ia supernovae have very uniform properties, and are useful standard candles over intergalactic distances. Some calibrations are required to compensate for the gradual change in properties or different frequencies of abnormal luminosity supernovae at high red shift, and for small variations in brightness identified by light curve shape or spectrum.

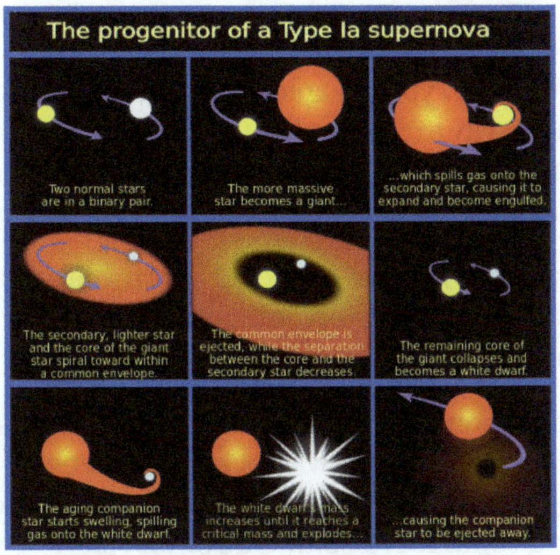

Formation of a type Ia supernova.

Normal Type Ia

There are several means by which a supernova of this type can form, but they share a common underlying mechanism. If a carbon-oxygen white dwarf accreted enough matter to reach the Chandrasekhar limit of about 1.44 solar masses ($M_\odot$) (for a non-rotating star), it would no longer be able to support the bulk of its mass through electron degeneracy pressure and would begin to collapse. However, the current view is that this limit is not normally attained; increasing temperature and density inside the core ignite carbon fusion as the star approaches the limit (to within about 1%), before collapse is initiated.

Within a few seconds, a substantial fraction of the matter in the white dwarf undergoes nuclear fusion, releasing enough energy (1–7044200000000000000♠2×10^44 J) to unbind the star in a supernova explosion. An outwardly expanding shock wave is generated, with matter reaching velocities on the order of 5,000–20,000 km/s, or roughly 3% of the speed of light. There is also a significant increase in luminosity, reaching an absolute magnitude of −19.3 (or 5 billion times brighter than the Sun), with little variation.

The model for the formation of this category of supernova is a closed binary star system. The larger of the two stars is the first to evolve off the main sequence, and it expands to form a red giant. The two stars now share a common envelope, causing their mutual orbit to shrink. The giant star then sheds most of its envelope, losing mass until it can no longer continue nuclear fusion. At this point it becomes a white dwarf star, composed primarily of carbon and oxygen. Eventually the secondary star also evolves off the main sequence to form a red giant. Matter from the giant is accreted by the white dwarf, causing the latter to increase in mass. Despite widespread acceptance of the basic model, the exact details of initiation and of the heavy elements produced in the explosion are still unclear.

Type Ia supernovae follow a characteristic light curve—the graph of luminosity as a function of time—after the explosion. This luminosity is generated by the radioactive decay of nickel-56 through cobalt-56 to iron-56. The peak luminosity of the light curve is extremely consistent across normal Type Ia supernovae, having a maximum absolute magnitude of about −19.3. This allows them to be used as a secondary standard candle to measure the distance to their host galaxies.

Non-Standard Type Ia

Another model for the formation of a Type Ia explosion involves the merger of two white dwarf stars, with the combined mass momentarily exceeding the Chandrasekhar limit. There is much variation in this type of explosion, and in many cases there may be no supernova at all, but it is expected that they will have a broader and less luminous light curve than the more normal Type Ia explosions.

Abnormally bright Type Ia supernovae are expected when the white dwarf already has a mass higher than the Chandrasekhar limit, possibly enhanced further by asymmetry, but the ejected material will have less than normal kinetic energy.

There is no formal sub-classification for the non-standard Type Ia supernovae. It has been proposed that a group of sub-luminous supernovae that occur when helium accretes onto a white dwarf should be classified as type Iax. This type of supernova may not always completely destroy the white dwarf progenitor and could leave behind a zombie star.

One specific type of non-standard Type Ia supernova develops hydrogen, and other, emission lines and gives the appearance of mixture between a normal Type Ia and a Type IIn supernova. Examples are SN 2002ic and SN 2005gj. These supernova have been dubbed Type Ia/IIn, Type Ian, Type IIa and Type IIan.

Core Collapse

Very massive stars can undergo core collapse when nuclear fusion suddenly becomes unable to sustain the core against its own gravity; this is the cause of all types of supernova except type Ia. The collapse may cause violent expulsion of the outer layers of the star resulting in a supernova, or the release of gravitational potential energy may be insufficient and the star may collapse into a black hole or neutron star with little radiated energy.

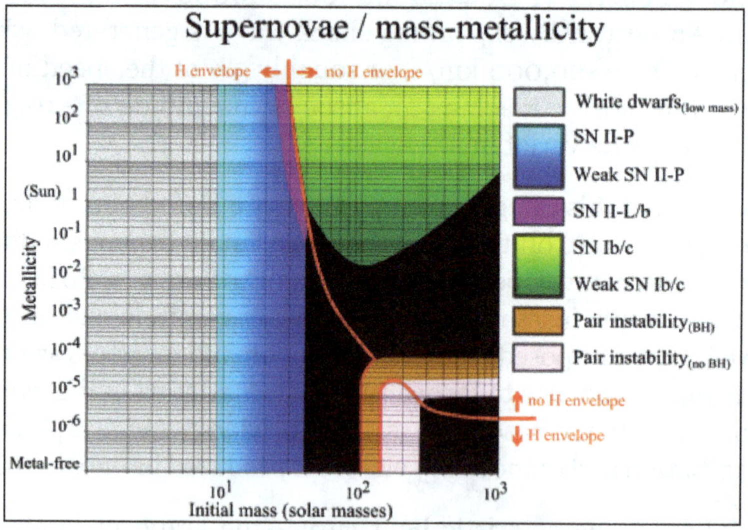

Supernova types by initial mass-metallicity.

Core collapse can be caused by several different mechanisms: electron capture; exceeding the Chandrasekhar limit; pair-instability; or photodisintegration. When a massive star develops an iron core larger than the Chandrasekhar mass it will no longer be able to support itself by electron degeneracy pressure and will collapse further to a neutron star or black hole. Electron capture by magnesium in a degenerate O/Ne/Mg core causes gravitational collapse followed by explosive oxygen fusion, with very similar results. Electron-positron pair production in a large post-helium burning core removes thermodynamic support and causes initial collapse followed by runaway fusion, resulting in a pair-instability supernova. A sufficiently large and hot stellar core may generate gamma-rays energetic enough to initiate photodisintegration directly, which will cause a complete collapse of the core.

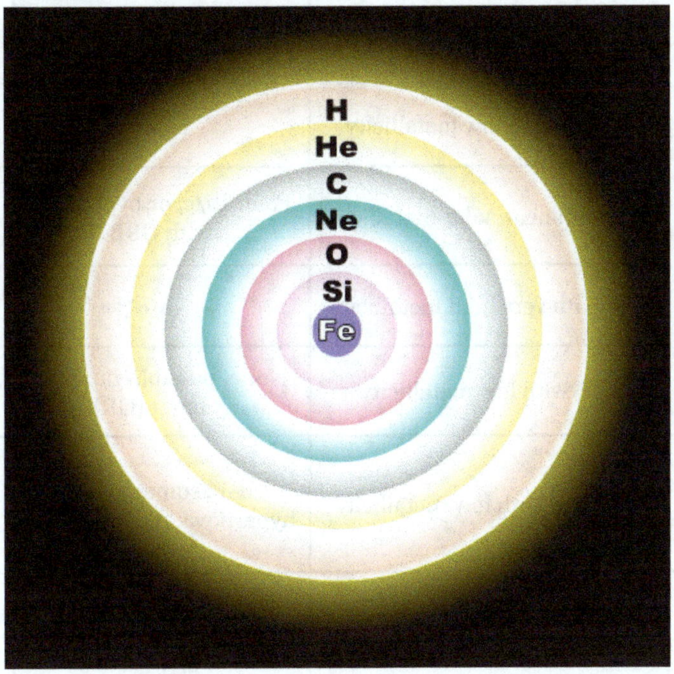

The onion-like layers of a massive, evolved star just prior to core collapse (Not to scale).

The table below lists the known reasons for core collapse in massive stars, the types of star that they occur in, their associated supernova type, and the remnant produced. The metallicity is the proportion of elements other than hydrogen or helium, as compared to the Sun. The initial mass is the mass of the star prior to the supernova event, given in multiples of the Sun's mass, although the mass at the time of the supernova may be much lower.

Type IIn supernovae are not listed in the table. They can potentially be produced by various types of core collapse in different progenitor stars, possibly even by type Ia white dwarf ignitions, although it seems that most will be from iron core collapse in luminous supergiants or hypergiants (including LBVs). The narrow spectral lines for which they are named occur because the supernova is expanding into a small dense cloud of circumstellar material. It appears that a significant proportion of supposed type IIn supernovae are actually supernova imposters, massive eruptions of LBV-like stars similar to the Great Eruption of Eta Carinae. In these events, material previously ejected from the star creates the narrow absorption lines and causes a shock wave through interaction with the newly ejected material.

Core collapse scenarios by mass and metallicity

Cause of collapse	Progenitor star approximate initial mass	Supernova Type	Remnant

Electron capture in a degenerate O+Ne+Mg core	8–10	Faint II-P	Neutron star
Iron core collapse	10–25	Faint II-P	Neutron star
	25–40 with low or solar metallicity	Normal II-P	Black hole after fallback of material onto an initial neutron star
	25–40 with very high metallicity	II-L or II-b	Neutron star
	40–90 with low metallicity	None	Black hole
	≥40 with near-solar metallicity	Faint Ib/c, or hypernova with GRB	Black hole after fallback of material onto an initial neutron star
	≥40 with very high metallicity	Ib/c	Neutron star
	≥90 with low metallicity	None, possible gamma-ray burst (GRB)	Black hole
Pair instability	140–250 with low metallicity	II-P, sometimes a hypernova, possible GRB	No remnant
Photodisintegration	≥250 with low metallicity	None (or luminous supernova?), possible GRB	Massive black hole

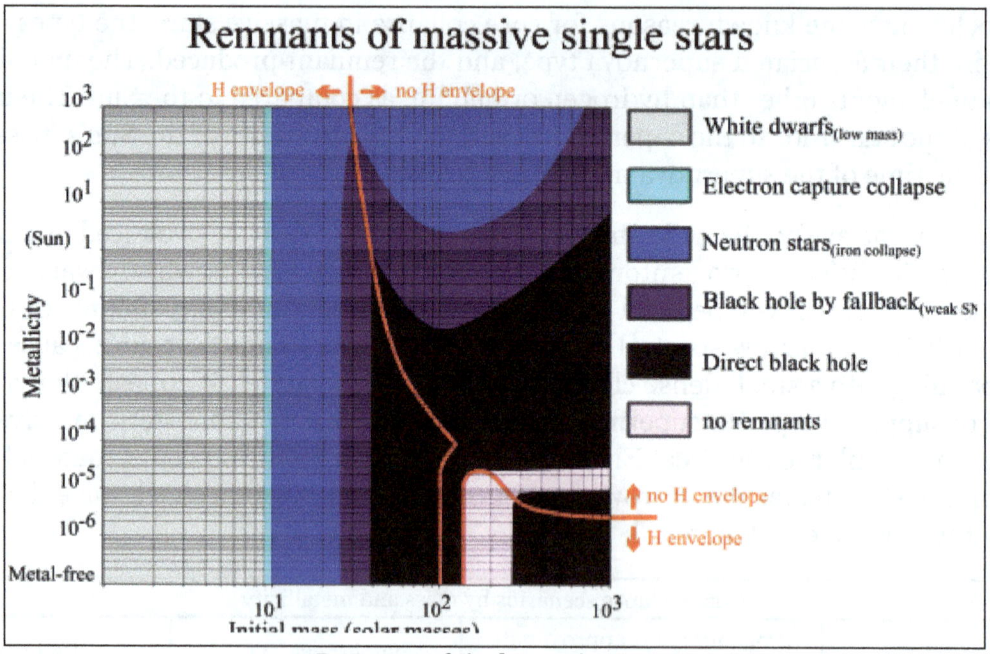

Remnants of single massive stars.

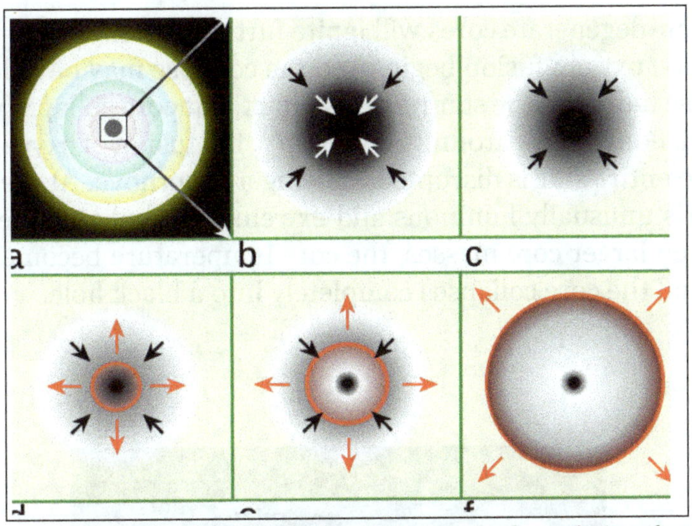

Within a massive, evolved star (a) the onion-layered shells of elements undergo fusion, forming an iron core (b) that reaches Chandrasekhar-mass and starts to collapse. The inner part of the core is compressed into neutrons (c), causing infalling material to bounce (d) and form an outward-propagating shock front (red). The shock starts to stall (e), but it is re-invigorated by a process that may include neutrino interaction. The surrounding material is blasted away (f), leaving only a degenerate remnant.

When a stellar core is no longer supported against gravity it collapses in on itself with velocities reaching 70,000 km/s (0.23c), resulting in a rapid increase in temperature and density. What follows next depends on the mass and structure of the collapsing core, with low mass degenerate cores forming neutron stars, higher mass degenerate cores mostly collapsing completely to black holes, and non-degenerate cores undergoing runaway fusion.

The initial collapse of degenerate cores is accelerated by beta decay, photodisintegration and electron capture, which causes a burst of electron neutrinos. As the density increases, neutrino emission is cut off as they become trapped in the core. The inner core eventually reaches typically 30 km diameter and a density comparable to that of an atomic nucleus, and neutron degeneracy pressure tries to halt the collapse. If the core mass is more than about 15 $M_{\odot}$ then neutron degeneracy is insufficient to stop the collapse and a black hole forms directly with no supernova explosion.

In lower mass cores the collapse is stopped and the newly formed neutron core has an initial temperature of about 100 billion kelvin, 6000 times the temperature of the sun's core. At this temperature, neutrino-antineutrino pairs of all flavors are efficiently formed by thermal emission. These thermal neutrinos are several times more abundant than the electron-capture neutrinos. About 10^{46} joules, approximately 10% of the star's rest mass, is converted into a ten-second burst of neutrinos which is the main output of the event. The suddenly halted core collapse rebounds and produces a shock wave that stalls within milliseconds in the outer core as energy is lost through the dissociation of heavy elements. A process that is not clearly understoodis necessary to allow the outer layers of the core to reabsorb around 10^{44} joules (1 foe) from the neutrino pulse, producing the visible explosion, although there are also other theories on how to power the explosion.

Some material from the outer envelope falls back onto the neutron star, and for cores beyond about 8 $M_{\odot}$ there is sufficient fallback to form a black hole. This fallback will reduce the kinetic energy of the explosion and the mass of expelled radioactive material, but in some situations it may also generate relativistic jets that result in a gamma-ray burst or an exceptionally luminous supernova.

Collapse of massive non-degenerate cores will ignite further fusion. When the core collapse is initiated by pair instability, oxygen fusion begins and the collapse may be halted. For core masses of 40–60 $M_{\odot}$, the collapse halts and the star remains intact, but core collapse will occur again when a larger core has formed. For cores of around 60–130 $M_{\odot}$, the fusion of oxygen and heavier elements is so energetic that the entire star is disrupted, causing a supernova. At the upper end of the mass range, the supernova is unusually luminous and extremely long-lived due to many solar masses of ejected ^{56}Ni. For even larger core masses, the core temperature becomes high enough to allow photodisintegration and the core collapses completely into a black hole.

Type II

The atypical subluminous type II SN 1997D.

Stars with initial masses less than about eight times the sun never develop a core large enough to collapse and they eventually lose their atmospheres to become white dwarfs. Stars with at least 9 $M_{\odot}$ (possibly as much as 12 $M_{\odot}$) evolve in a complex fashion, progressively burning heavier elements at hotter temperatures in their cores. The star becomes layered like an onion, with the burning of more easily fused elements occurring in larger shells. Although popularly described as an onion with an iron core, the least massive supernova progenitors only have oxygen-neon(-magnesium) cores. These super AGB stars may form the majority of core collapse supernovae, although less luminous and so less commonly observed than those from more massive progenitors.

If core collapse occurs during a supergiant phase when the star still has a hydrogen envelope, the result is a Type II supernova. The rate of mass loss for luminous stars depends on the metallicity and luminosity. Extremely luminous stars at near solar metallicity will lose all their hydrogen before they reach core collapse and so will not form a type II supernova. At low metallicity, all stars will reach core collapse with a hydrogen envelope but sufficiently massive stars collapse directly to a black hole without producing a visible supernova.

Stars with an initial mass up to about 90 times the sun, or a little less at high metallicity, are expected to result in a type II-P supernova which is the most commonly observed type. At moderate to high metallicity, stars near the upper end of that mass range will have lost most of their hydrogen when core collapse occurs and the result will be a Type II-L supernova. At very low metallicity, stars of around 140–250 $M_{\odot}$ will reach core collapse by pair instability while they still have a hydrogen atmosphere and an oxygen core and the result will be a supernova with Type II characteristics but a very large mass of ejected Ni and high luminosity.

Type Ib and Ic

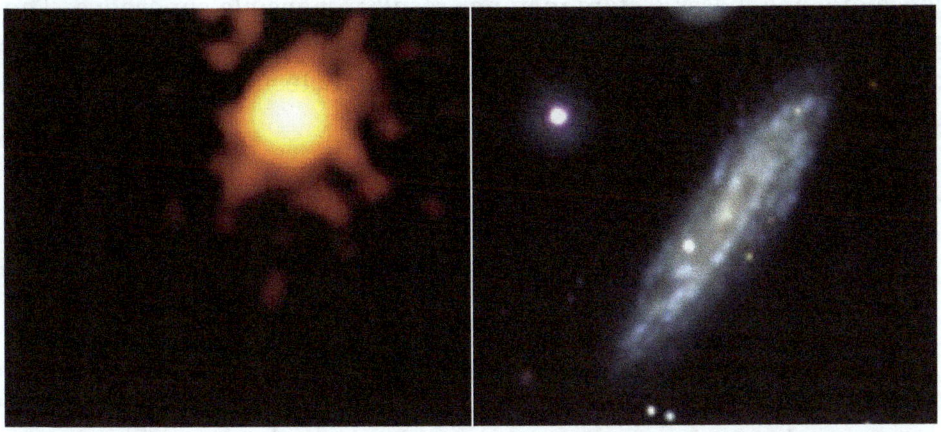

SN 2008D, a Type Ib supernova, shown in X-ray (left) and visible light (right) at the far upper end of the galaxy.

These supernovae, like those of Type II, are massive stars that undergo core collapse. However the stars which become Types Ib and Ic supernovae have lost most of their outer (hydrogen) envelopes due to strong stellar winds or else from interaction with a companion. These stars are known as Wolf-Rayet stars, and they occur at moderate to high metallicity where continuum driven winds cause sufficiently high mass loss rates. Observations of type Ib/c supernova do not match the observed or expected occurrence of Wolf Rayet stars and alternate explanations for this type of core collapse supernova involve stars stripped of their hydrogen by binary interactions. Binary models provide a better match for the observed supernovae, with the proviso that no suitable binary helium stars have ever been observed. Since a supernova explosion can occur whenever the mass of the star at the time of core collapse is low enough not to cause complete fallback to a black hole, any massive star may result in a supernova if it loses enough mass before core collapse occurs.

Type Ib supernovae are the more common and result from Wolf-Rayet stars of type WC which still have helium in their atmospheres. For a narrow range of masses, stars evolve further before reaching core collapse to become WO stars with very little helium remaining and these are the progenitors of type Ic supernovae.

A few percent of the Type Ic supernovae are associated with gamma-ray bursts (GRB), though it is also believed that any hydrogen-stripped Type Ib or Ic supernova could produce a GRB, depending on the geometry of the explosion. The mechanism for producing this type of GRB is the jets produced by the magnetic field of the rapidly spinning magnetar formed at the collapsing core of the star. The jets would also transfer energy into the expanding outer shell of the explosion to produce a super-luminous supernova.

Ultra-stripped supernovae occur when the exploding star has been stripped (almost) all the way to the metal core, via mass transfer in a close binary. As a result, very little material is ejected from the exploding star (~0.1 M_{Sun}). In the most extreme cases, ultra-stripped supernovae can occur in naked metal cores, barely above the Chandrasekhar mass limit. SN 2005ek might be an observational example of an ultra-stripped supernova, giving rise to a relatively dim and fast decaying light curve. The nature of ultra-stripped supernovae can be both iron core-collapse and electron capture supernovae, depending on the mass of the collapsing core.

Failed

The core collapse of some massive stars may not result in a visible supernova. The main model for this is a sufficiently massive core that the explosion is insufficient to reverse the infall of the outer layers onto a black hole. These events are difficult to detect, but large surveys have detected possible candidates.

Light Curves

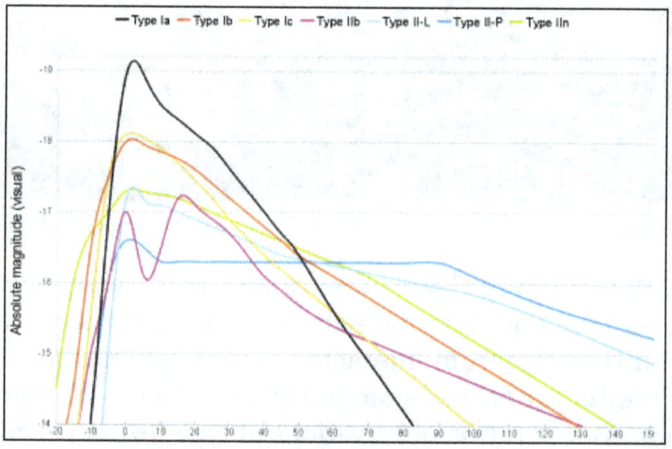

Comparative supernova type light curves.

A historic puzzle concerned the source of energy that can maintain the optical supernova glow for months. Although the energy that disrupts each type of supernovae is delivered promptly, the light curves are mostly dominated by subsequent radioactive heating of the rapidly expanding ejecta. Some have considered rotational energy from the central pulsar. The ejecta gases would dim quickly without some energy input to keep it hot. The intensely radioactive nature of the ejecta gases, which is now known to be correct for most supernovae, was first calculated on sound nucleosynthesis grounds in the late 1960s. It was not until SN 1987A that direct observation of gamma-ray lines unambiguously identified the major radioactive nuclei.

It is now known by direct observation that much of the light curve (the graph of luminosity as a function of time) after the explosion of a Type II Supernova such as SN 1987A is provided its energy by those predicted radioactive decays. Although the luminous emission consists of optical photons, it is the radioactive power absorbed by the ejected gases that keeps the remnant hot enough to radiate light. The radioactive decay of through its daughters to produces gamma-ray photons , primarily of 847keV and 1238keV, that are absorbed and dominate the heating and thus the luminosity of the ejecta at intermediate times (several weeks) to late times (several months).

Energy for the peak of the light curve of SN1987A was provided by the decay of Ni to Co (half life 6 days) while energy for the later light curve in particular fit very closely with the 77.3 day half-life of Co decaying to Fe. Later measurements by space gamma-ray telescopes of the small fraction of the Co and Co gamma rays that escaped the SN 1987A remnant without absorption confirmed earlier predictions that those two radioactive nuclei were the power sources.

The visual light curves of the different supernova types all depend at late times on radioactive heating, but they vary in shape and amplitude on the underlying mechanisms of the explosion, the way that visible radiation is produced, the epoch of its observation, and the transparency of the ejected material. The light curves can be significantly different at other wavelengths. For example, at ultraviolet wavelengths there is an early extremely luminous peak lasting only a few hours corresponding to the breakout of the shock launched by the initial explosion, but that breakout is hardly detectable optically.

The light curves for type Ia are mostly very uniform, with a consistent maximum absolute magnitude and a relatively steep decline in luminosity. Their optical energy output is driven by radioactive decay of nickel-56 (half life 6 days), which then decays to radioactive cobalt-56 (half life 77 days). These radioisotopes from material ejected in the explosion excite surrounding material to incandescence. Studies of cosmology today rely on radioactivity providing the energy for the optical brightness of supernovae of Type Ia, which are the "standard candles" of cosmology but whose diagnostic 847keV and 1238keV gamma rays were first detected only in 2014. The initial phases of the light curve decline steeply as the effective size of the photosphere decreases and trapped electromagnetic radiation is depleted. The light curve continues to decline in the B band while it may show a small shoulder in the visual at about 40 days, but this is only a hint of a secondary maximum that occurs in the infra-red as certain ionised heavy elements recombine to produce infra-red radiation and the ejecta become transparent to it. The visual light curve continues to decline at a rate slightly greater than the decay rate of the radioactive cobalt (which has the longer half life and controls the later curve), because the ejected material becomes more diffuse and less able to convert the high energy radiation into visual radiation. After several months, the light curve changes its decline rate again as positron emission becomes dominant from the remaining cobalt-56, although this portion of the light curve has been little-studied.

Type Ib and Ic light curves are basically similar to type Ia although with a lower average peak luminosity. The visual light output is again due to radioactive decay being converted into visual radiation, but there is a much lower mass of nickel-56 produced in these types of explosion. The peak luminosity varies considerably and there are even occasional type Ib/c supernovae orders of magnitude more and less luminous than the norm. The most luminous type Ic supernovae are referred to as hypernovae and tend to have broadened light curves in addition to the increased peak luminosity. The source of the extra energy is thought to be relativistic jets driven by the formation of a rotating black hole, which also produce gamma-ray bursts.

The light curves for type II supernovae are characterised by a much slower decline than type I, on the order of 0.05 magnitudes per day, excluding the plateau phase. The visual light output is dominated by kinetic energy rather than radioactive decay for several months, due primarily to the existence of hydrogen in the ejecta from the atmosphere of the supergiant progenitor star. In the initial explosion this hydrogen becomes heated and ionised. The majority of type II supernovae show a prolonged plateau in their light curves as this hydrogen recombines, emitting visible light and becoming more transparent. This is then followed by a declining light curve driven by radioactive decay although slower than in type I supernovae, due to the efficiency of conversion into light by all the hydrogen.

In type II-L the plateau is absent because the progenitor had relatively little hydrogen left in its atmosphere, sufficient to appear in the spectrum but insufficient to produce a noticeable plateau in the light output. In type IIb supernovae the hydrogen atmosphere of the progenitor is so depleted (thought to be due to tidal stripping by a companion star) that the light curve is closer to a type I supernova and the hydrogen even disappears from the spectrum after several weeks.

Type IIn supernovae are characterised by additional narrow spectral lines produced in a dense shell of circumstellar material. Their light curves are generally very broad and extended, occasionally also extremely luminous and referred to as a hypernova. These light curves are produced by the highly efficient conversion of kinetic energy of the ejecta into electromagnetic radiation by interaction with the dense shell of material. This only occurs when the material is sufficiently dense and compact, indicating that it has been produced by the progenitor star itself only shortly before the supernova occurs.

Large numbers of supernovae have been catalogued and classified to provide distance candles and test models. Average characteristics vary somewhat with distance and type of host galaxy, but can broadly be specified for each supernova type.

Physical properties of supernovae by type				
Type[a]	Average peak absolute magnitude[b]	Approximate energy (foe)[c]	Days to peak luminosity	Days from peak to 10% luminosity
Ia	−19	1	approx. 19	around 60
Ib/c (faint)	around −15	0.1	15–25	unknown
Ib	around −17	1	15–25	40–100
Ic	around −16	1	15–25	40–100
Ic (bright)	to −22	above 5	roughly 25	roughly 100
II-b	around −17	1	around 20	around 100
II-L	around −17	1	around 13	around 150
II-P (faint)	around −14	0.1	roughly 15	unknown
II-P	around −16	1	around 15	Plateau then around 50
IIn[d]	around −17	1	12–30 or more	50–150
IIn (bright)	to −22	above 5	above 50	above 100

Asymmetry

The pulsar in the Crab nebula is travelling at 375 km/s relative to the nebula.

A long-standing puzzle surrounding Type II supernovae is why the compact object remaining after the explosion is given a large velocity away from the epicentre; pulsars, and thus neutron stars, are observed to have high velocities, and black holes presumably do as well, although they are far harder to observe in isolation. The initial impetus can be substantial, propelling an object of more than a solar mass at a velocity of 500 km/s or greater. This indicates an asymmetry in the explosion, but the mechanism by which momentum is transferred to the compact object remainsa puzzle. Proposed explanations for this kick include convection in the collapsing star and jet production during neutron star formation.

One possible explanation for the asymmetry in the explosion is large-scale convection above the core. The convection can create variations in the local abundances of elements, resulting in uneven nuclear burning during the collapse, bounce and resulting explosion.

Another possible explanation is that accretion of gas onto the central neutron star can create a disk that drives highly directional jets, propelling matter at a high velocity out of the star, and driving transverse shocks that completely disrupt the star. These jets might play a crucial role in the resulting supernova explosion. (A similar model is now favored for explaining long gamma-ray bursts.)

Initial asymmetries have also been confirmed in Type Ia supernova explosions through observation. This result may mean that the initial luminosity of this type of supernova depends on the viewing angle. However, the explosion becomes more symmetrical with the passage of time. Early asymmetries are detectable by measuring the polarization of the emitted light.

Energy Output

Although we are used to thinking of supernovae primarily as luminous visible events, the electromagnetic radiation they release is almost a minor side-effect of the explosion. Particularly in the case of core collapse supernovae, the emitted electromagnetic radiation is a tiny fraction of the total energy released during the event.

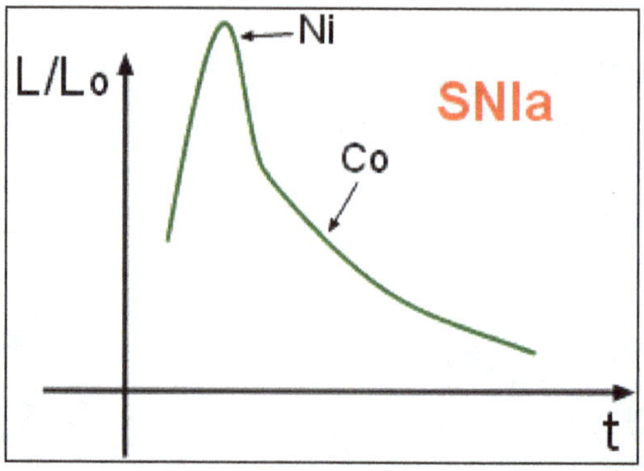

The radioactive decays of nickel-56 and cobalt-56 that produce a supernova visible light curve.

There is a fundamental difference between the balance of energy production in the different types of supernova. In type Ia white dwarf detonations, most of the energy is directed into heavy element synthesis and the kinetic energy of the ejecta. In core collapse supernovae, the vast majority of the energy is directed into neutrino emission, and while some of this apparently powers the following main explosion 99%+ of the neutrinos escape the star in the first few minutes following the start of the collapse.

Type Ia supernovae derive their energy from a runaway nuclear fusion of a carbon-oxygen white dwarf. The details of the energetics are still not fully understood, but the end result is the ejection of the entire mass of the original star at high kinetic energy. Around half a solar mass of that mass is Ni generated from silicon burning. Ni is radioactive and decays into Co by beta plus decay (with a half life of six days) and gamma rays. Co itself decays by the beta plus (an anti-electron) path with a half life of 77 days into stable Fe. These two processes are responsible for the electromagnetic radiation from type Ia supernovae. In combination with the changing transparency of the ejected material, they produce the rapidly declining light curve.

Core collapse supernovae are on average visually fainter than type Ia supernovae, but the total energy released is far higher. In the case of supernovae, the energy of gravitational potential energy is converted into kinetic energy that compresses and collapses the core, initially producing electron neutrinos from disintegrating nucleons, followed by all flavours of thermal neutrinos from the super-heated neutron star core. Around 1% of these neutrinos are thought to deposit sufficient energy into the outer layers of the star to drive the resulting explosion, but again the details cannot be reproduced exactly in current models. Kinetic energies and nickel yields are somewhat lower than type Ia supernovae, hence the lower visual luminosity of supernovae, but energy from the de-ionisation of the many solar masses of remaining hydrogen can contribute to a much slower decline in luminosity and produce the plateau phase seen in the majority of core collapse supernovae.

Energetics of supernovae					
Supernova	Approximate total energy 10^{44} joules (foe)[c]	Ejected Ni (solar masses)	Neutrino energy (foe)	Kinetic energy (foe)	Electromagnetic radiation (foe)
Type Ia	1.5	0.4 – 0.8	0.1	1.3 – 1.4	~0.01

Core collapse	100	(0.01) – 1	100	1	0.001 – 0.01
Hypernova	100	~1	1–100	1–100	~0.1
Pair instability	5–100	0.5 – 50	low?	1–100	0.01 – 0.1

In some core collapse supernovae, fallback onto a black hole drives relativistic jets which may produce a brief energetic and directional burst of gamma rays and also transfers substantial further energy into the ejected material. This is one scenario for producing high luminosity supernovae and is thought to be the cause of type Ic hypernovae and long duration gamma-ray bursts. If the relativistic jets are too brief and fail to penetrate the stellar envelope then a low luminosity gamma-ray burst may be produced and the supernova may be sub-luminous.

When a supernova occurs inside a small dense cloud of circumstellar material, it will produce a shock wave that can efficiently convert a high fraction of the kinetic energy into electromagnetic radiation. Even though the initial explosion energy was entirely normal the resulting supernova will have high luminosity and extended duration since it does not rely on exponential radioactive decay. This type of event may cause type IIn hypernovae.

Although pair-instability supernovae are core collapse supernovae with spectra and light curves similar to type II-P, the nature of that explosion following core collapse is more like that of a giant type Ia with runaway fusion of carbon, oxygen, and silicon. The total energy released by the highest mass events is comparable to other core collapse supernovae but neutrino production is thought to be very low, hence the kinetic and electromagnetic energy released is very high. The cores of these stars are much larger than any white dwarf and the amount of radioactive nickel and other heavy elements ejected from their cores can be orders of magnitude higher, with consequently high visual luminosity.

Progenitor

The supernova classification type is closely tied to the type of star at the time of the explosion. The occurrence of each type of supernova depends dramatically on the metallicity and hence the age of the host galaxy.

Shown in this sped-up artist's impression, is a collection of distant galaxies, the occasional supernova can be seen. Each of these exploding stars briefly rivals the brightness of its host galaxy.

Type Ia supernovae are produced from white dwarf stars in binary systems and occur in all galaxy types. Core collapse supernovae are only found in galaxies undergoing current or very recent star formation, since they result from short-lived massive stars. They are most commonly found in type Sc spirals, but also in the arms of other spiral galaxies and in irregular galaxies, especially starburst galaxies.

Type Ib/c and II-L, and possibly most type IIn, supernovae are only thought to be produced from stars having near-solar metallicity levels that result in high mass loss from massive stars, hence they are less common in older more distant galaxies. The table shows the expected progenitor for the main types of core collapse supernova, and the approximate proportions of each in the local neighbourhood.

Fraction of core collapse supernovae types by progenitor		
Type	Progenitor star	Fraction
Ib	WC Wolf-Rayet	9%
Ic	WO Wolf-Rayet	17%
II-P	Supergiant	55%
II-L	Supergiant with a depleted hydrogen shell	3.0%
IIn	Supergiant in a dense cloud of expelled material (such as LBV)	2.4%
IIb	Supergiant with highly depleted hydrogen (stripped by companion?)	12%
IIpec	Blue supergiant?	1.0%

There are a number of difficulties reconciling modelled and observed stellar evolution leading up to core collapse supernovae. Red supergiants are the expected progenitors for the vast majority of core collapse supernovae, and these have been observed but only at relatively low masses and luminosities, below about 18 $M_\odot$ and 100,000 $L_\odot$ respectively. Most progenitors of type II supernovae are not detected and must be considerably fainter, and presumably less massive. It is now proposed that higher mass red supergiants do not explode as supernovae, but instead evolve back towards hotter temperatures. Several progenitors of type IIb supernovae have been confirmed, and these were K and G supergiants, plus one A supergiant. Yellow hypergiants or LBVs are proposed progenitors for type IIb supernovae, and almost all type IIb supernovae near enough to observe have shown such progenitors.

Until just a few decades ago, hot supergiants were not considered likely to explode, but observations have shown otherwise. Blue supergiants form an unexpectedly high proportion of confirmed supernova progenitors, partly due to their high luminosity and easy detection, while not a single Wolf-Rayet progenitor has yet been clearly identified. Models have had difficulty showing how blue supergiants lose enough mass to reach supernova without progressing to a different evolutionary stage. One study has shown a possible route for low-luminosity post-red supergiant luminous blue variables to collapse, most likely as a type IIn supernova.

The expected progenitors of type Ib supernovae, luminous WC stars, are not observed at all. Instead WC stars are found at lower luminosities, apparently post-red supergiant stars. WO stars are extremely rare and visually relatively faint, so it is difficult to say whether such progenitors are missing or just yet to be observed. Very luminous progenitors, despite numerous supernovae being observed near enough that such progenitors would have been clearly imaged. Several examples of hot luminous progenitors of type IIn supernovae have been detected: SN 2005gy and SN 2010jl were both apparently massive luminous stars, but are very distant; and SN 2009ip had a highly luminous progenitor likely to have been an LBV, but is a peculiar supernova whose exact nature is disupted.

Interstellar Impact

Source of Heavy Elements

Supernovae are the major source of elements heavier than oxygen. These elements are produced by nuclear fusion for nuclei up to S, by silicon photodisintegration rearrangement and quasiequilibrium during silicon burning for nuclei between Ar and Ni, and by rapid captures of neutrons during the supernova explosion for elements heavier than iron. Nucleosynthesis during silicon burning yields nuclei roughly 1000-100,000 times more abundant than the r-process isotopes heavier than iron. Supernovae are the most likely, although not undisputed, candidate sites for the r-process, which is the rapid capture of neutrons that occurs at high temperature and high density of neutrons. Those reactions produce highly unstable nuclei that are rich in neutrons and that rapidly beta decay into more stable forms. The r-process produces about half of all the heavier isotopes of the elements beyond iron, including plutonium and uranium. The only other major competing process for producing elements heavier than iron is the s-process in large, old, red-giant AGB stars, which produces these elements slowly over longer epochs and which cannot produce elements heavier than lead.

Role in Stellar Evolution

The remnant of a supernova explosion consists of a compact object and a rapidly expanding shock wave of material. This cloud of material sweeps up the surrounding interstellar medium during a free expansion phase, which can last for up to two centuries. The wave then gradually undergoes a period of adiabatic expansion, and will slowly cool and mix with the surrounding interstellar medium over a period of about 10,000 years.

Supernova remnant N 63A lies within a clumpy region of gas and dust in the Large Magellanic Cloud.

The Big Bang produced hydrogen, helium, and traces of lithium, while all heavier elements are synthesized in stars and supernovae. Supernovae tend to enrich the surrounding interstellar medium with *metals*—elements other than hydrogen and helium.

These injected elements ultimately enrich the molecular clouds that are the sites of star formation. Thus, each stellar generation has a slightly different composition, going from an almost pure mixture of hydrogen and helium to a more metal-rich composition. Supernovae are the dominant mechanism for distributing these heavier elements, which are formed in a star during its period of nuclear fusion. The different abundances of elements in the material that forms a star have important influences on the star's life, and may decisively influence the possibility of having planets orbiting it.

The kinetic energy of an expanding supernova remnant can trigger star formation due to compression of nearby, dense molecular clouds in space. The increase in turbulent pressure can also prevent star formation if the cloud is unable to lose the excess energy.

Evidence from daughter products of short-lived radioactive isotopes shows that a nearby supernova helped determine the composition of the Solar System 4.5 billion years ago, and may even have triggered the formation of this system. Supernova production of heavy elements over astronomic periods of time ultimately made the chemistry of life on Earth possible.

Effect on Earth

A near-Earth supernova is a supernova close enough to the Earth to have noticeable effects on its biosphere. Depending upon the type and energy of the supernova, it could be as far as 3000 light-years away. Gamma rays from a supernova would induce a chemical reaction in the upper atmosphere converting molecular nitrogen into nitrogen oxides, depleting the ozone layer enough to expose the surface to harmful solar radiation. This has been proposed as the cause of the Ordovician–Silurian extinction, which resulted in the death of nearly 60% of the oceanic life on Earth. In 1996 it was theorized that traces of past supernovae might be detectable on Earth in the form of metal isotope signatures in rock strata. Iron-60 enrichment was later reported in deep-sea rock of the Pacific Ocean. In 2009, elevated levels of nitrate ions were found in Antarctic ice, which coincided with the 1006 and 1054 supernovae. Gamma rays from these supernovae could have boosted levels of nitrogen oxides, which became trapped in the ice.

Type Ia supernovae are thought to be potentially the most dangerous if they occur close enough to the Earth. Because these supernovae arise from dim, common white dwarf stars in binary systems, it is likely that a supernova that can affect the Earth will occur unpredictably and in a star system that is not well studied. The closest known candidate is IK Pegasi. Recent estimates predict that a Type II supernova would have to be closer than eight parsecs (26 light-years) to destroy half of the Earth's ozone layer, and there are no such candidates closer than about 500 light years.

Milky Way Candidates

The next supernova in the Milky Way will likely be detectable even if it occurs on the far side of the galaxy. It is likely to be produced by the collapse of an unremarkable red supergiant and it is very probable that it will already have been catalogued in infrared surveys such as 2MASS. There is a smaller chance that the next core collapse supernova will be produced by a different type of massive star such as a yellow hypergiant, luminous blue variable, or Wolf-Rayet. The chances of the next supernova being a type Ia produced by a white dwarf are calculated to be about a third of those for a core collapse supernova. Again it should be observable wherever it occurs, but it is less likely that the progenitor will ever have been observed prior to the explosion. It isn't even known exactly what a type Ia progenitor system looks like, and difficult to detect them beyond a few parsecs. The total supernova rate in our galaxy is estimated to be about 4.6 per century, or one every 22 years, although we haven't actually observed one for several centuries.

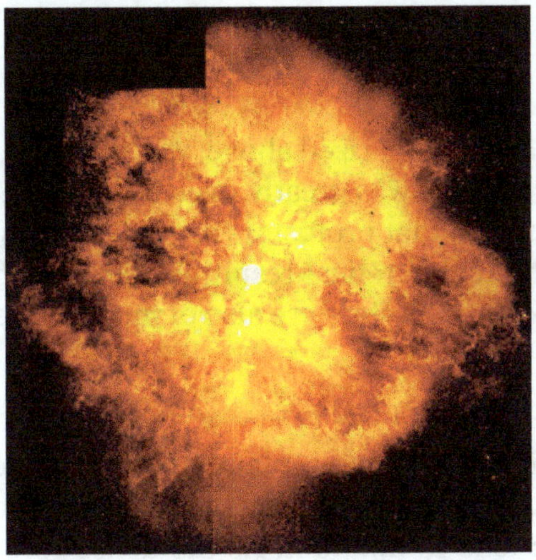

The nebula around Wolf–Rayet star WR124, which is located at a distance of about 21,000 light years.

Statistically, the next supernova is likely to be produced from an otherwise unremarkable red supergiant, but it is difficult to identify which of those supergiants are in the final stages of heavy element fusion in their cores and which have millions of years left. The most massive red supergiants are expected to shed their atmospheres and evolve to Wolf-Rayet stars before their cores collapse. All Wolf-Rayet stars are expected to end their lives from the Wolf-Rayet phase within a million years or so, but again it is difficult to identify those that are closest to core collapse. One class that is expected to have no more than a few thousand years before exploding are the WO Wolf-Rayet stars, which are known to have exhausted their core helium. Only eight of them are known, and only four of those are in the Milky Way.

A number of close or well known stars have been identified as possible core collapse supernova candidates: the red supergiants Antares and Betelgeuze; the yellow hypergiant Rho Cassiopeiae; the luminous blue variable Eta Carinae that has already produced a supernova imposter explosion; and the brightest component, a Wolf–Rayet star in the Regor or Gamma Velorum system, Others have gained notoriety as possible, although not very likely, progenitors for a gamma-ray burst; for example WR 104.

Identification of candidates for a type Ia supernova explosion is much more speculative. Any binary with an accreting white dwarf might produce a supernova although the exact mechanism and timescale is still debated. These systems are faint and difficult to identify, but the novae and recurrent novae are such systems that conveniently advertise themselves. One examples is U Scorpii, The nearest known type Ia supernova candidate is IK Pegasi (HR 8210), located at a distance of 150 light-years, but observations suggest it will be several million years before the white dwarf can accrete the critical mass required to become a Type Ia supernova.

Protostar

A protostar is a very young star that is still gathering mass from its parent molecular cloud. The protostellar phase is the earliest one in the process of stellar evolution. For a one solar-mass star it lasts about 1,000,000 years. The phase begins when a molecular cloud first collapses under the force of self-gravity. It ends when the protostar blows back the infalling gas and is revealed as an optically visible pre-main-sequence star, which later contracts to become a main sequence star.

History

The modern picture of protostars, summarized above, was first suggested by Chushiro Hayashi. In the first models, the size of protostars was greatly overestimated. Subsequent numerical calculations clarified the issue, and showed that protostars are only modestly larger than main-sequence stars of the same mass. This basic theoretical result has been confirmed by observations, which find that the largest pre-main-sequence stars are also of modest size.

Protostellar Evolution

Star formation begins in relatively small molecular clouds called dense cores. Each dense core is initially in balance between self-gravity, which tends to compress the object, and both gas pressure and magnetic pressure, which tend to inflate it. As the dense core accrues mass from its larger, surrounding cloud, self-gravity begins to overwhelm pressure, and collapse begins. Theoretical modeling of an idealized spherical cloud initially supported only by gas pressure indicates that the collapse process spreads from the inside toward the outside. Spectroscopic observations of dense cores that do not yet contain stars indicate that contraction indeed occurs. So far, however, the predicted outward spread of the collapse region has not been observed.

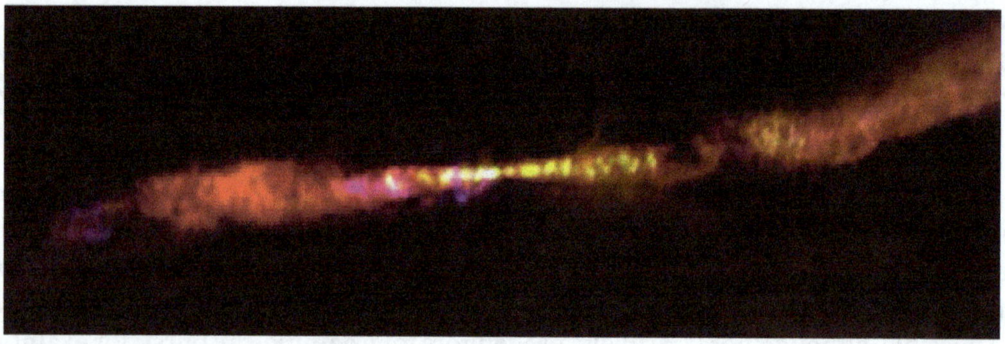

Infant star CARMA-7 and its jets are located approximately 1400 light-years from Earth within the Serpens South star cluster.

The gas that collapses toward the center of the dense core first builds up a low-mass protostar, and then a protoplanetary disk orbiting the object. As the collapse continues, an increasing amount of gas impacts the disk rather than the star, a consequence of angular momentum conservation. Exactly how material in the disk spirals inward onto the protostar is not yet understood, despite a great deal of theoretical effort. This problem is illustrative of the larger issue of accretion disk theory, which plays a role in much of astrophysics.

HBC 1 is a young pre-main-sequence star.

Regardless of the details, the outer surface of a protostar consists at least partially of shocked gas that has fallen from the inner edge of the disk. The surface is thus very different from the relatively quiescent photosphere of a pre-main sequence or main-sequence star. Within its deep interior, the protostar has lower temperature than an ordinary star. At its center, hydrogen is not yet undergoing nuclear fusion. Theory predicts, however, that the hydrogen isotope deuterium is undergoing fusion, creating helium-3. The heat from this fusion reaction tends to inflate the protostar, and thereby helps determine the size of the youngest observed pre-main-sequence stars.

The energy generated from ordinary stars comes from the nuclear fusion occurring at their centers. Protostars also generate energy, but it comes from the radiation liberated at the shocks on its surface and on the surface of its surrounding disk. The radiation thus created must traverse the interstellar dust in the surrounding dense core. The dust absorbs all impinging photons and reradiates them at longer wavelengths. Consequently, a protostar is not detectable at optical wavelengths, and cannot be placed in the Hertzsprung-Russell diagram, unlike the more evolved pre-main-sequence stars.

The actual radiation emanating from a protostar is predicted to be in the infrared and millimeter regimes. Point-like sources of such long-wavelength radiation are commonly seen in regions that are obscured by molecular clouds. It is commonly believed that those conventionally labeled as Class 0 or Class I sources are protostars. However, there is still no definitive evidence for this identification.

Chemical Composition

The Sun as a protostar had the same composition as today, which is 71.1% hydrogen, 27.4% helium, and 1.5% heavier elements, by mass.

Observed Classes of Young Stars

For details of observational classification, see Table below.

Class	peak emission	duration (Years)
0	submillimeter	10^4
I	far-infrared	10^5
II	near-infrared	10^6
III	visible	10^7

References

- Plait, Philip C. (2002). Bad astronomy: misconceptions and misuses revealed, from astrology to the moon landing "hoax". John Wiley and Sons. pp. 237–240. ISBN 0-471-40976-6
- Alves, J.; Lada, C.; Lada, E. (2001). Tracing H_2 Via Infrared Dust Extinction. Cambridge University Press. p. 217. ISBN 0-521-78224-4
- Ballesteros-Paredes, J.; Klessen, R. S.; Mac Low, M.-M.; Vazquez-Semadeni, E. "Molecular Cloud Turbulence and Star Formation". In Reipurth, B.; Jewitt, D.; Keil, K. Protostars and Planets V. pp. 63–80. ISBN 0-8165-2654-0
- Lyall, Francis; Larsen, Paul B. (2009). "Chapter 7: The Moon and Other Celestial Bodies". Space Law: A Treatise. Ashgate Publishing, Ltd. p. 176. ISBN 0-7546-4390-5
- Murdin, P. (November 2000). "Aristillus (c. 200 BC)". Encyclopedia of Astronomy and Astrophysics. Bibcode:2000eaa..bookE3440.. doi:10.1888/0333750888/3440. ISBN 0-333-75088-8

6

Solar Astrophysics

Solar astrophysics is a branch of astrophysics that specializes in the study of the sun. A detailed understanding of the sun is crucial before contemplating the formation and evolution of the Solar System, the Ring System, the Planetary System and other similar alien systems which maybe discovered in the near future. This chapter provides a complete and detailed examination of solar astrophysics for a clearer understanding.

The Solar system (or solar system) is the home stellar system for human beings and all known forms of life. The solar system comprises the Sun, all the objects gravitationally bound to it, and the heliosphere, an enormous magnetic bubble enclosing most of the known solar system, including the solar wind and the entire solar magnetic field. Objects bound gravitationally to the Sun include eight planets, their 165 known moons, at least three dwarf planets (including Pluto) with their four known moons, and billions of small bodies, classified primarily as asteroids, meteoroids, or comets.

In broad terms, the solar system comprises regions starting with the Sun and extending outward from there successively to:

- Four terrestrial inner planets.
- The asteroid belt of small rocky bodies.
- Four gas giant outer planets.
- The Kuiper belt of icy objects.
- The scattered disc.
- The heliopause.
- The hypothetical Oort cloud.

The solar system major bodies drawn to scale.

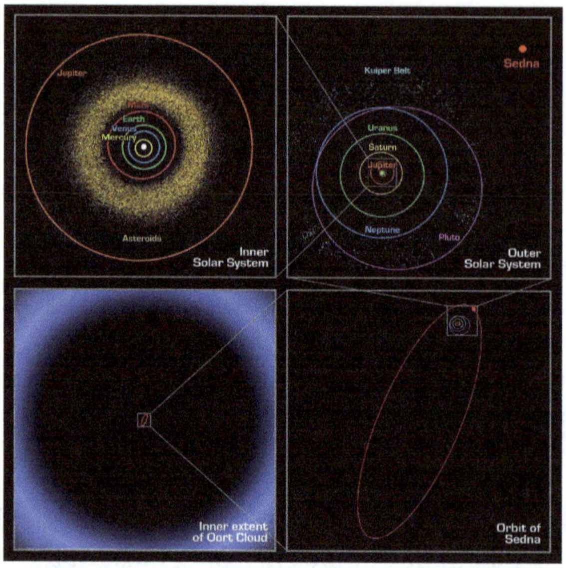

The solar system major orbits drawn to scale (clockwise from top left)

All solar system objects are immersed in either the solar plasma, known as the solar wind, the interstellar plasma of the Milky Way galaxy, or a mixture of the two in a region lying just inside the heliopause, the diaphonous boundary of the heliosphere.

The eight planets, in order of their distances from the Sun, are: Mercury, Venus, Earth, Mars, Jupiter, Saturn, Uranus, and Neptune. Six of the eight planets are in turn orbited by natural satellites, usually termed "moons" after Earth's Moon, and each of the outer planets is encircled by planetary rings of dust and other particles. All of the planets, including the earth are surrounded by a teardrop shaped magnetosphere generated by the interaction of the magnetic field of the solar wind with that of each planet. All the planets except Earth are named after gods and goddesses from Greco-Roman mythology.

The solar system's position in the Milky Way galaxy and the Earth's position in the solar system are remarkably well-suited for the emergence and sustenance of life on Earth. The solar system occupies the Milky Way galaxy's relatively narrow "habitable zone" where the intensity of gamma rays, x-rays, and other radiation from surrounding stars is not too great to block the development of life by sterilizing planetary surfaces. This zone is comfortably removed from the intense concentration of stars closer to the galactic centre, but not too close to the outer margins of the galaxy where there are not enough of the heavier elements to support formation of planets. Within the solar system, the Earth is neither so close to the sun as to be too hot for life nor so far away as to be too cold. Had the Earth been much closer to the Sun or much farther away, life would not have arisen on the planet.

Historically, the solar system has been the key performer in the greatest and grandest show on earth, the drama of the sky with the sun moving across it each day and with the fixed stars of the constellations providing each night a slowly moving backdrop against which the more agile moon, and five visible planets Mercury, Venus, Mars, Jupiter, and Saturn performed their routines. Massive monuments such as Stonehenge in England and pyramids in Egypt and the Americas attest to ancient peoples' profound efforts to observe and predict movements of the solar bodies, and the active fields today of both astrology and solar astronomy demonstrate the continued importance of the solar system in human life. In western cultures, the names of the days of the week are derived from names of the planets or the names of gods associated with them. The current solar system model with the powerful central sun guiding from afar the planets and other solid bodies and radiating the light and heat that sustain life itself embodies an archetype that has been adopted into other fields.

Major features of the solar system (not to scale): The Sun, the eight planets, the asteroid belt containing the dwarf planet Ceres, the dwarf planet Pluto (in outermost orbit), and a comet (the dwarf planet Eris is not shown).

The principal component of the solar system is the Sun (astronomical symbol ☉); a main sequence G2 star that contains 99.86 percent of the system's known mass and dominates it gravitationally. Because of its large mass, the Sun has an interior density high enough to sustain nuclear fusion, releasing enormous amounts of energy, most of which is radiated into space in the form of electromagnetic radiation, including visible light. Jupiter and Saturn, the two largest bodies orbiting the Sun, account for more than 90 percent of the system's remaining mass (the postulated Oort cloud, should its existence be confirmed, would also hold a substantial percentage).

The ecliptic viewed in sunlight from behind the Moon in this image taken by the Clementine spacecraft. From left to right: Mercury, Mars, Saturn.

Most objects in orbit around the Sun lie within the same shallow plane, called the *ecliptic*, which is roughly parallel to the Sun's equator. The planets lie very close to the ecliptic, while comets and

Kuiper belt objects often lie at significant angles to it. All of the planets and most other objects orbit with the Sun's rotation in a counterclockwise direction as viewed from a point above the Sun's north pole. There is a direct relationship between how far away a planet is from the Sun and how quickly it orbits (Kepler's laws of planetary motion). Mercury, with the smallest orbital circumference, travels the fastest, while Neptune, being much farther from the Sun, travels slower.

A planet's distance from the Sun varies in the course of its year. Its closest approach to the Sun is known as its *perihelion*, while its farthest point from the Sun is called its *aphelion*. Though planets follow nearly circular orbits, with perihelions roughly equal to their aphelions; many comets, asteroids and objects of the Kuiper belt follow highly elliptical orbits, with large differences between perihelion and aphelion.

Astronomers most often measure distances within the solar system in astronomical units (AU). One AU is the average distance between the Earth and the Sun, or roughly 149,598,000 kilometers (93 million miles). Pluto is roughly 38 AU from the Sun, while Jupiter lies at roughly 5.2 AU.

Representing the planets and the sun as a unit solar system in a single graphic presents major challenges because both the individual bodies (diameters ranging from 5,000 to 1,390,000 kilometers) and the planets' distances from the sun (average orbital radii ranging from .4 to 38 AU) are so greatly varied. As a result, many representations of the solar system distort the planetary orbits, with equal spacing between each object. In actuality, the general pattern is that the farther a planet or a belt is from the Sun, the greater the distance between it and the previous orbit. Attempts have been made to find a correlation between these distances (an early one was Bode's Law), but to date there is no accepted theory that explains the respective orbital distances.

Classification of Solar System Bodies

A planet, as defined by the IAU General Assembly on August 24, 2006, is any body in orbit around the Sun that (a) has enough mass to form itself into a near-spherical shape, and (b) has cleared or captured its immediate neighbourhood of all smaller objects. Eight objects in the solar system currently meet these criteria: Mercury, Venus, Earth, Mars, Jupiter, Saturn, Uranus, and Neptune.

The IAU also used the term dwarf planet, a newly defined class of solar system bodies. The key difference between planets and dwarf planets is that while both are required to orbit the Sun and be of large enough mass that their own gravity pulls them into a nearly round shape, dwarf planets are not required to "clear their neighbourhood" of other celestial bodies. Three objects in the solar system are currently included in this category. They are: Pluto, the largest known Kuiper belt object; Ceres, the largest object in the asteroid belt; and the scattered disc object Eris. The IAU is evaluating other known objects to see if they fit within the definition of dwarf planets. The most likely candidates are some of the larger asteroids and several trans-Neptunian objects.

The remaining objects in the solar system were classified as small solar system bodies (SSSB). The IAU used the term SSSB to describe solar system objects that are neither planets nor dwarf planets:

> *All other objects orbiting the Sun shall be referred to collectively as "Small Solar System Bodies" These currently include most of the Solar System asteroids, most Trans-Neptunian Objects (TNOs), comets, and other small bodies.*

As of 2006, the IAU considers the following bodies to be SSSBs:

1. All asteroids except Ceres.
2. All centaurs.
3. All trans-Neptunian objects, including Kuiper belt and scattered disc objects, with the exception of Pluto and Eris.
4. All comets.

Sun

The Sun as seen from Earth.

The Sun is the solar system's parent star, and far and away its chief component. It is classified as a moderately large yellow dwarf. This name, however, is misleading, because on the scale of stars in our galaxy, the Sun is rather large and bright. Stars are classified based on their position on the Hertzsprung-Russell diagram, a graph that plots the brightness of stars against their surface temperature. Generally speaking, the hotter a star is, the brighter it is. Stars that follow this pattern are said to be on the "main sequence," and the Sun lies right in the middle of it. This has led many astronomy textbooks to label the Sun as "average," although stars brighter and hotter than it are rare, whereas dimmer and cooler stars are common. The vast majority of stars are dim red dwarfs, though they are under-represented in star catalogs because we can observe only those that are relatively close to the Sun.

According to current theories of stellar evolution, the Sun's position on the main sequence means that it is in the "prime of life" for a star. In other words, it has not yet exhausted its store of hydrogen for nuclear fusion, and thus it has not been forced (as older red giants must) to fuse more inefficient elements such as helium and carbon. As it ages, the Sun is growing increasingly bright. Early in its history, it was roughly 75 percent as bright as it is today. Calculations of the ratios of hydrogen and helium within the Sun suggest it is roughly halfway through its life cycle and will eventually begin moving off the main sequence becoming larger, brighter, and redder, and becoming a red giant in about five billion years.

The Sun is a "population I" star meaning that it is fairly new in galactic terms, having been born in the later stages of the universe's evolution. As such, it contains far more elements heavier than hydrogen and helium (metals, in astronomical parlance) than older, population II stars, such as those found in globular star clusters. Because elements heavier than hydrogen and helium were formed in the cores of ancient and exploding stars, the first generation of stars had to die before the universe could be enriched with those elements. For this reason, the very oldest stars contain very little "metal," while stars born later have more. This high "metallicity" is thought to have been crucial for the Sun to develop a planetary system, because planets form from the accretion of metals.

The Sun ejects the solar wind, a pulsing stream of electrons, ions, neutral atoms, and magnetic fields, at speeds greater than 2 million kilometers per hour. It thus creates a very tenuous "atmosphere" (the heliosphere), that permeates the solar system for at least 100 AU. This environment is known as the interplanetary medium. Small quantities of cosmic dust (some of it arguably interstellar in origin) are also present in the interplanetary medium and are responsible for the phenomenon known as zodiacal light. The influence of the Sun's rotating magnetic field on the interplanetary medium creates one of the largest structures in the solar system: the heliospheric current sheet.

Inner Planets

The inner planets, from left to right: Mercury, Venus, Earth, and Mars (sizes to scale).

The four inner (or terrestrial) planets are characterized by their dense, rocky composition, few or no moons, and lack of ring systems. They are composed largely of minerals with high melting points, such as silicates, that form their solid crusts and semi-liquid mantles. In addition, metallic dust grains such as iron form the cores of these planets. Three of the four inner planets have atmospheres. All have impact craters, and all but one possess tectonic surface features, such as rift valleys and volcanoes. The term *inner planet* should not be confused with *inferior planet*, which designates planets that are closer to the Sun than the Earth is (namely, Mercury and Venus).

Mercury

Mercury (0.4 AU), the closest planet to the Sun, is also the least massive of the planets, at only 0.055 of Earth's mass. Mercury has a very thin atmosphere, consisting of atoms blasted off its surface by the solar wind. Mercury is so hot that these atoms quickly escape into space. Thus, in contrast to the Earth and Venus, which have stable atmospheres, Mercury's atmosphere is constantly being replenished. Mercury is surrounded by extremely small amounts of helium, hydrogen, oxygen, and sodium. This envelope of gases is so thin that the greatest possible atmospheric pressure (force exerted by the weight of gases) on Mercury would be about 2×10^{-12} kg/cm^2 (3×10^{-11} psi). The atmospheric pressure on Earth is about 1.03 kg/cm^2 (14.7 psi). It has no natural satellites, and, to date, no observed geological activity other than that produced by impacts. Its relatively large iron core and thin mantle have not yet been adequately explained. Some have postulated that its outer layers may have been stripped off by a giant impact, and that it was prevented from fully accreting by the Sun's gravity. The MESSENGER spacecraft should aid in resolving this issue when it arrives in Mercury's orbit in 2011.

Venus

Venus (0.7 AU), the first truly terrestrial planet, is of mass comparable to that of the Earth (0.815 of Earth mass) and has been called "Earth's twin." Like Earth, it possesses a substantial atmosphere, a thick silicate mantle around an iron core, and evidence of one-time internal geological activity, such as volcanoes. It is, however, much drier than Earth, its atmosphere is 90 times as dense, and it is composed overwhelmingly of carbon dioxide and sulfuric acid. Evidence suggests that the crust of Venus, unlike Earth's crust, is not divided into tectonic plates, but rather comprises a single, thick rind. Venus has no natural satellite. Also, it is the hottest planet, despite being farther from the Sun than Mercury, with temperatures reaching more than 400 °C. This is most likely due to the large amount of greenhouse gases (gases that trap heat) in the atmosphere.

Earth

The largest and densest of the inner planets, Earth (1 AU) is the only planet known to have life. It is also the only one to demonstrate unequivocal evidence of current geological activity. Its liquid hydrosphere, unique among the terrestrial planets, is probably the reason Earth is also the only planet where multi-plate tectonics has been observed because water acts as a lubricant for subduction.

The Earth's atmosphere is radically different from that of the other terrestrial planets, having been altered by the presence of life to contain 21 percent free oxygen. Its satellite, the Moon, is unusually large compared with the size of the Earth, causing significant effects on Earth such as ocean tides. In addition, the relative sizes and distances of the Moon and Sun enable us to observe a total solar eclipse. Earth's magnetic field protects its atmosphere from interacting with the solar wind, except in the vicinity of the poles where a variable influx of charged particles from the solar wind at times generates the shimmering Northern and Southern Lights.

Mars

Mars (1.5 AU), at only 0.107 of Earth's mass, is less massive than either Earth or Venus. It possesses a tenuous atmosphere of carbon dioxide. Its surface, peppered with vast volcanoes and rift valleys such as Valles Marineris, shows that it was once geologically active and recent evidence suggests this may have continued until relatively recently. Mars possesses two tiny moons, Deimos and Phobos, thought to be captured asteroids. Evidences from Martian meteorites, spacecraft observations, and rover experiments suggest that Mars may have had liquid water in the past, and some think that Mars may have harbored some primitive forms of life as well.

Asteroid Belt

Asteroids are mostly small solar system bodies that are composed in significant part of rocky, nonvolatile minerals. The main asteroid belt occupies an orbit between Mars and Jupiter, between 2.3 and 3.3 AU from the Sun. It is thought to consist of pieces that, due to the gravitational interference of Jupiter, failed to coalesce into a small terrestrial planet.

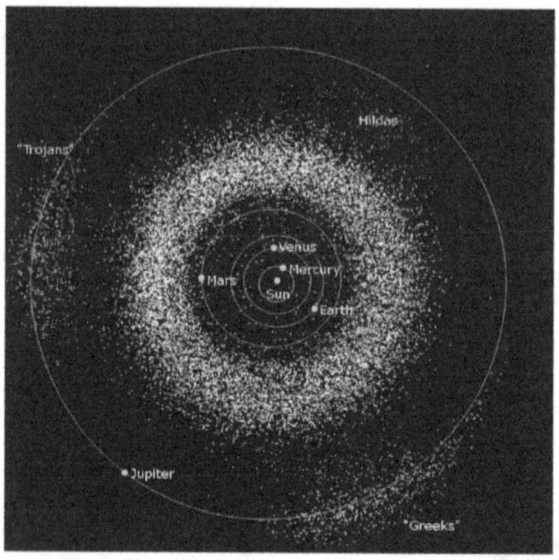

Image of the main asteroid belt and the Trojan asteroids.

Asteroids range in size from hundreds of kilometers in diameter down to dust particles. All asteroids save the largest, Ceres, are classified as small solar system bodies. A number of other asteroids such as Vesta and Hygeia could potentially be reclassified as dwarf planets if it can be conclusively shown that they are spherical. The asteroid belt contains tens of thousands and potentially millions of objects over one kilometer in diameter. Despite their large numbers, the total mass of the main belt is unlikely to be more than a thousandth of the Earth's mass. In contrast to its various depictions in science fiction, the main belt is very sparsely populated spacecraft routinely pass through it without incident. Asteroids with a diameter of less than 50 meters are called meteoroids.

Ceres

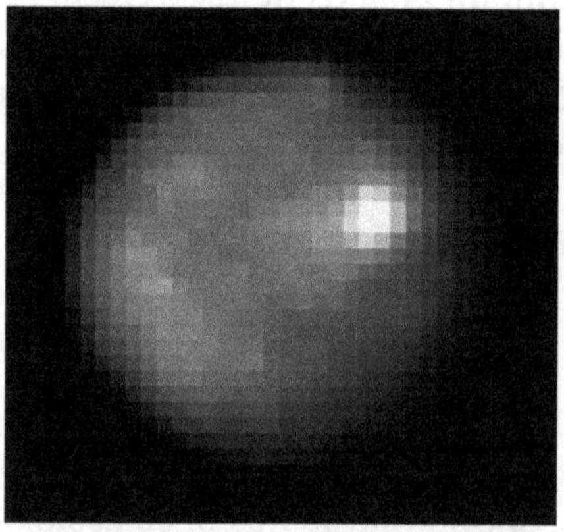

Ceres, as imaged by the Hubble Space Telescope.

Ceres (2.77 AU) is the largest astronomical body in the asteroid belt and the only known dwarf planet in this region. It has a diameter of slightly under 1,000 kilometers large enough for its own gravity to pull it into a spherical shape. Ceres was considered a planet when it was discovered in the nineteenth century, but it was reclassified as an asteroid when further observations revealed additional asteroids. Its location coincides with what is predicted by Bode's Law.

Asteroid Groups

Asteroids in the main belt are subdivided into asteroid groups and families, based on their specific orbital characteristics. Asteroid moons are asteroids that orbit larger asteroids. They are not as clearly distinguished as planetary moons, sometimes being almost as large as their partners. The asteroid belt also contains main-belt comets, which may have been the source of Earth's water.

Trojan asteroids are located in either of Jupiter's "Lagrange points," which are gravitationally-stable regions leading and trailing the planet in its orbit. The term "Trojan asteroids" is also sometimes used for asteroids in the Lagrange points of other planets as well.

The inner solar system is also dusted with rogue asteroids, many of which cross the orbits of the inner planets.

Outer Planets

From top to bottom: Neptune, Uranus, Saturn, and Jupiter (sizes not to scale).

The four outer planets, or gas giants, (sometimes called Jovian planets) are so large that they collectively make up 99 percent of the mass known to orbit the Sun. Jupiter and Saturn are true giants, at 318 and 95 Earth masses, respectively, and composed largely of hydrogen and helium. Uranus and Neptune are substantially smaller, being only 14 and 17 Earth masses, respectively. Their atmospheres contain a smaller percentage of hydrogen and helium, and a higher percentage of "ices," such as frozen water, ammonia, and methane. For this reason some astronomers have suggested putting them in a separate category"Uranian planets" or "ice giants." All four outer planets exhibit orbital debris rings, although only the ring system of Saturn is easily observable from Earth. The term *outer planet* should not be confused with *superior planet*, which designates all planets that lie outside Earth's orbit (thus consisting of the outer planets plus Mars).

Jupiter

Jupiter (5.2 AU), at 318 Earth masses, is 2.5 times the mass of all the other planets put together. Its composition of largely hydrogen and helium is not very different from that of the Sun. Jupiter's strong internal heat creates a number of semi-permanent features in its atmosphere, such as cloud bands and the Great Red Spot. Three of its 63 satellites Ganymede, Io, and Europa share elements in common with the terrestrial planets, such as volcanism and internal heating. Ganymede has a larger diameter than Mercury.

Saturn

Saturn (9.5 AU), famous for its extensive ring system, has many qualities in common with Jupiter, including its atmospheric composition, though it is far less massive, being only 95 Earth masses. Two of its 49 moons, Titan and Enceladus, show signs of geological activity, though they are largely made of ice. Titan, like Ganymede, is larger than Mercury; it is also the only satellite in the solar system with a substantial atmosphere.

Uranus

Uranus (19.6 AU), at 14 Earth masses, is the lightest of the outer planets. Uniquely among the planets, it orbits the Sun on its side; its axial tilt lies at over 90 degrees to the ecliptic. Its core is remarkably cooler than that of the other gas giants (though it is still several thousand degrees Celsius) and radiates very little heat into space. Uranus has 27 satellites, the largest being Titania, Oberon, Umbriel, Ariel, and Miranda.

Neptune

Neptune (30 AU), though slightly smaller than Uranus, is denser and slightly more massive, at 17 Earth masses. It radiates more internal heat than Uranus, but not as much as Jupiter or Saturn. Its peculiar ring system is composed of a number of dense "arcs" of material separated by gaps. Neptune has 13 moons. The largest, Triton, is geologically active, with geysers of liquid nitrogen.

Edgeworth-kuiper Belt

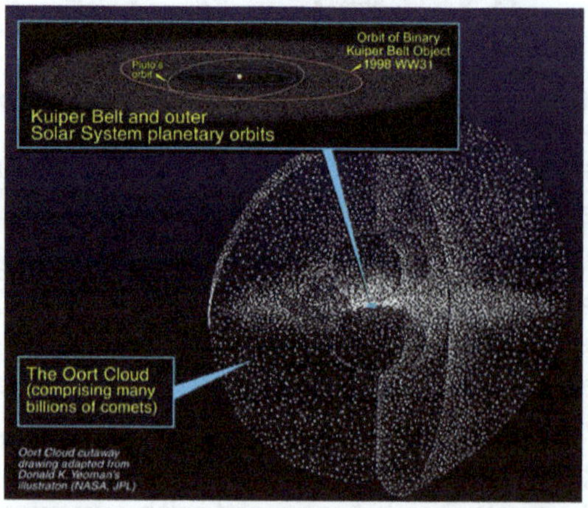

Artist's rendering of the Kuiper Belt and hypothetical Oort cloud.

The area beyond Neptune, sometimes referred to as the "outer solar system" or the "trans-Neptunian region," is still largely unexplored.

This region's first formation, which actually begins inside Neptune's orbit, is called the Edgeworth-Kuiper belt (or simply, Kuiper belt). This belt is a great ring of debris similar to the asteroid belt, but it is composed mainly of ice and is far greater in extent, lying between 30 and 50 AU from the Sun. This region is thought to be the place of origin for short-period comets, such as Halley's comet. Though it is composed mainly of small solar system bodies, many of the largest Kuiper belt objects could soon be reclassified as dwarf planets. There are estimated to be over 100,000 Kuiper belt objects with a diameter greater than 50 kilometers; however, the total mass of the Kuiper belt is relatively low, perhaps barely equaling the Earth's mass. Many Kuiper belt objects have multiple satellites, and most have orbits that take them outside the plane of the ecliptic.

The Kuiper belt can be roughly divided into two regions:

1. The "resonant" belt, consisting of objects whose orbits are in some way linked to that of Neptune (orbiting, for instance, three times for every two Neptune orbits, or twice for every one). It actually begins within the orbit of Neptune itself.

2. The "classical" belt, consisting of objects that don't have any resonance with Neptune. It extends from roughly 39.4 AU to 47.7 AU.

Pluto and Charon

Pluto, Charon, and their two known moons.

From 1930 to 2006, Pluto (39 AU average), one of the largest known Kuiper belt objects, was considered the solar system's ninth planet. In 2006, however, the International Astronomical Union (IAU) created an official definition of a planet that disqualified Pluto as a planet. As a result, Pluto was reclassified as a dwarf planet. Pluto has a relatively eccentric orbit, inclined 17 degrees to the ecliptic plane and ranging from 29.7 AU from the Sun at perihelion (within the orbit of Neptune) to 49.5 AU at aphelion. Prior to the 2006 redefinitions, Charon was considered a moon of Pluto, but in light of the redefinition, it is unclear whether Charon will continue to be classified as Pluto's moon or as a dwarf planet itself. Charon does not orbit Pluto rather, both bodies orbit a barycentre of gravity in empty space, making Pluto-Charon a binary system. Two much smaller moons, Nix and Hydra, orbit Pluto and Charon.

Kuiper belt objects that, like Pluto, possess a 3:2 orbital resonance with Neptune that is, they orbit twice for every three Neptunian orbits are called Plutinos. Other Kuiper belt objects have different resonant orbits (such as 2:1, 4:7, 3:5) and are grouped accordingly. The remaining Kuiper belt objects, in more "classical" orbits, are classified as Cubewanos.

Comets

Comets are small solar system bodies (usually only a few kilometers across) containing large amounts of volatile ice and tracing highly eccentric orbits. They generally have a perihelion within the orbit of the inner planets and an aphelion far beyond Pluto. When a comet approaches the Sun, its icy surface begins to sublimate, or boil away, creating a coma a long tail of gas and dust that is often visible with the naked eye.

Comet Hale-Bopp.

There are two basic types of comets:

1. Short-period comets, with orbits less than two hundred years.
2. Long-period comets, with orbits lasting thousands of years.

Short-period comets are believed to originate in the Kuiper belt, while long period comets, such as Hale-Bopp (pictured), are thought to originate in the Oort cloud. Some comets with hyperbolic orbits may originate outside the solar system. Old comets that have had most of their volatiles driven out by solar warming are often categorized as asteroids (likely to be D-type asteroids).

Centaurs are icy, comet-like bodies that have less eccentric orbits, so that they remain in the region between Jupiter and Neptune. The first centaur to be discovered, 2060 Chiron, has been called a comet because it was found to develop a coma just as comets do when they approach the Sun.

Scattered Disc

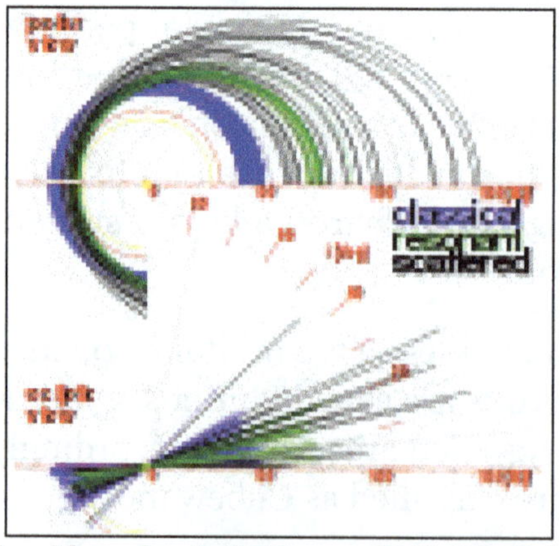

Black: scattered disc; blue: classical Edgeworth-Kuiper belt; green: resonant Edgeworth-Kuiper belt objects, including Pluto.

Overlapping the Kuiper belt but extending much further outward is what is called the scattered disc. Scattered disc objects are believed to have been originally native to the Kuiper belt but were ejected into erratic orbits in the outer fringes by the gravitational influence of Neptune's outward migration. Most scattered disc objects have perihelia within the Kuiper belt but aphelia as far as 150 AU from the Sun. Their orbits are also highly inclined to the ecliptic plane, and are often almost perpendicular to it. Some astronomers, such as Kuiper belt co-discoverer David Jewitt, consider the scattered disc to be merely another region of the Kuiper belt, and they describe scattered disc objects as "scattered Kuiper belt objects."

Eris

Eris (68 AU average) is the largest known scattered disc object. It was the cause of the most recent debate about what constitutes a planet, because it is at least five percent larger than Pluto, with an estimated diameter of 2,400 kilometers (1,500 miles). It is now the largest of the known dwarf planets. It has one moon named Dysnomia.

This object has many similarities with Pluto. Its orbit is highly eccentric, with a perihelion of 38.2 AU (roughly Pluto's distance from the Sun) and an aphelion of 97.6 AU. Also, the orbit is steeply inclined to the ecliptic plane, at 44 degrees, more so than any known object in the solar system except for the newly discovered object 2004 XR_{190}(also known as "Buffy"). It is believed to consist largely of rock and ice.

Farthest Regions

The point at which the solar system ends and interstellar space begins is not precisely defined, because the solar system's outer boundaries are delineated by two separate forces: the solar wind and the Sun's gravity. The solar wind extends to a point roughly 130 AU from the Sun, whereupon it surrenders to the surrounding environment of the interstellar medium. It is generally accepted, however, that the Sun's gravity holds sway over the hypothetical Oort cloud. This cloud, conceived of as a great mass of up to a trillion icy objects, is believed to be the source of all long-period comets. It is thought to surround the solar system like a shell from 50,000 to 100,000 AU beyond the Sun, or almost a quarter of the distance to the next star system. From this perspective, the vast majority of the solar system is completely unknown. Nonetheless, recent observations of the Solar system and other star systems have led to an increased understanding of what is or may be lying at its outer edge.

Sedna

An artist's conception of Sedna.

Sedna is a large, reddish Pluto-like object with a gigantic, highly elliptical orbit that takes it from about 76 AU at perihelion to 928 AU at aphelion and takes 12,050 years to complete. Mike Brown, who discovered the object in 2003, asserts that it cannot be part of the scattered disc or Kuiper Belt, as it has too distant a perihelion to have been affected by Neptune's migration. He and other astronomers consider it to be the first in an entirely new population one that may include the object 2000 CR_{105}, which has a perihelion of 45 AU, an aphelion of 415 AU, and an orbital period of 3,420 years. Sedna is very likely a dwarf planet, though its shape has yet to be determined with certainty.

Heliopause

The solar wind, a stream of ions, electrons, neutral matter, and magnetic fields emanating in great pulses from the sun, sustains the heliosphere, which is thought to be slightly teardrop shaped and to extend out from the sun to about 95 AU, or two and one half times the orbit of Pluto. The bubble's diaphonous edge begins at the termination shock the point at which the solar wind is significantly slowed by collision with opposing plasma winds of the interstellar medium. Here the wind condenses and becomes more turbulent in a region known as the heliosheath, which extends outward for a further 40 AU at its stellar-windward side, but stretches many times that distance in the opposite direction. The outer boundary of the sheath, the heliopause, is the region in which the solar wind is fully overcome by the interstellar galactic wind. Beyond the heliopause, at around 230 AU, the interstellar plasma registers the approach of the solar system in the form of the bow shock a plasma "wake" generated by the Sun as it travels through the Milky Way.

Galactic Context

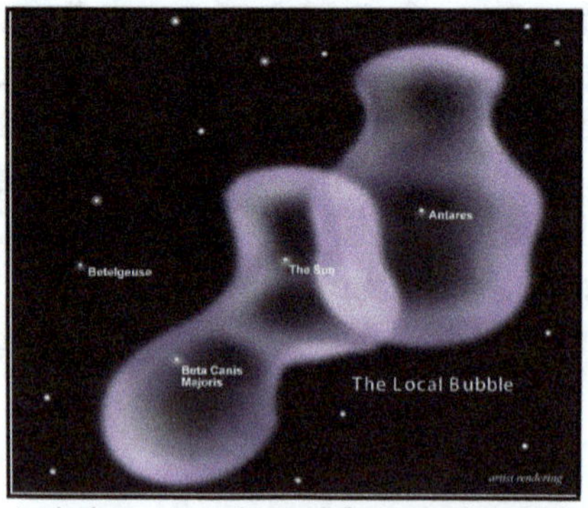

Artist's conception of the Local Bubble.

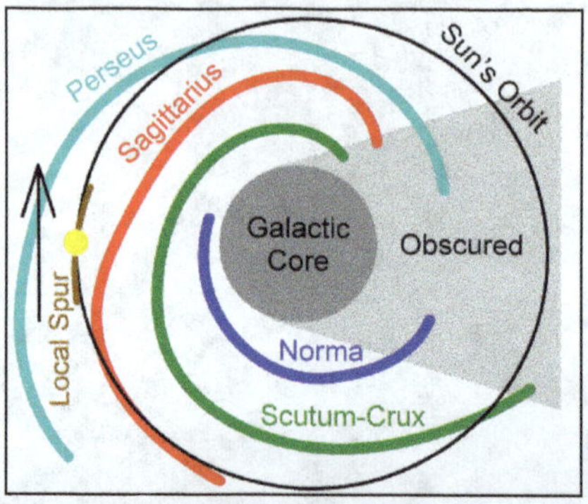

Presumed location of the Solar system within our galaxy.

The solar system is located in the Milky Way galaxy, a barred spiral galaxy with a diameter estimated at about 100,000 light years containing approximately 200 billion stars. Our Sun resides in one of the Milky Way's outer spiral arms, known as the Orion Arm or Local Spur. The immediate galactic neighbourhood of the solar system is known as the Local Fluff, an area of dense cloud in an otherwise sparse region known as the Local Bubble, an hourglass-shaped cavity in the interstellar medium, roughly three hundred light years across. The bubble is suffused with high-temperature plasma that suggests it is the product of several recent supernovae.

Estimates place the solar system at between 25,000 and 28,000 light-years from the galactic centre. Its speed is about 220 kilometers per second, and it completes one revolution every 226 million years. It appears to have a very remarkable orbit one that is both extremely close to being circular, and at nearly the exact distance at which the orbital speed matches the speed of the compression waves that form the spiral arms.

The apex of solar motion that is, the direction in which the Sun is heading is near the current location of the bright star, Vega. At the galactic location of the solar system, the escape velocity with regard to the gravity of the Milky Way is about 1,000 kilometers per second.

Extrasolar Planets

Until 1994, the solar system was the only known example of a stellar system in which a planet or planets orbit around a star. Since that first discovery, the subsequent discovery of more than 200 extrasolar planets has led to the term "solar system" being applied generically to all stellar systems. Technically, however, it should strictly refer to Earth's system only, as the word "solar" is derived from the Sun's Latin name, *Sol*. Other stellar systems or planetary systems are usually referred to by the names of their parent star, such as "the Alpha Centauri system" or "the 51 Pegasi system."

Studies of extrasolar systems have shown that the vast majority discovered so far are markedly different from the solar system. For instance, many extrasolar planetary systems contain a "hot Jupiter" a planet of size comparable to Jupiter, that nonetheless orbits very close to its star (for instance, at 0.05 AU). It has been hypothesized that while the giant planets in these systems probably formed in the same relative place as the gas giants in Earth's solar system did, the extrasolar giant planets may have migrated inward toward their parent star and along the way caused the destruction or ejection from the system of any existing smaller planets. The apparent prevalence of hot Jupiter's among the extrasolar planets discovered so far may be an artifact of a skewed data set arising simply because close-in giant planets are the easiest extrasolar planets to detect. In coming years, it is quite possible that ongoing improvements in the capabilities for detecting extrasolar planets will lead to the discovery of stellar systems much more similar to our own.

Discovery and Exploration

For many thousands of years, most people thought of the Earth as a stationary body at the centre of the universe, and categorically different from the ethereal objects that moved through the sky. The conceptual advances of the seventeenth century led by Nicolaus Copernicus, Galileo Galilei, Johannes Kepler, and Isaac Newton gradually guided people to accept ideas such as (a) the Earth moves in an orbit around the Sun, (b) the planets are governed by the same laws as the Earth, and (c) the planets could therefore be similar to the Earth.

Telescopic Observations

When astronomers first began using the telescope, they made unexpected discoveries about the solar system. For instance, Galileo Galilei discovered that the Moon was cratered, that the Sun was pocked with sunspots, and that Jupiter had four satellites in orbit around it. Giovanni Domenico Cassini and Christian Huygens followed on from Galileo's discoveries by discovering the rings of Saturn, and Saturn's moon Titan, respectively.

In 1682 Edmund Halley realized that repeated sightings of a comet were in fact recording the same object, returning regularly once every 75-76 years. This proved that comets were not atmospheric phenomena, as was previously thought, and provided the first evidence that objects other than planets orbited the Sun.

In 1781 William Herschel was looking for binary stars in the constellation of Taurus when he observed what he thought was a new comet. In fact, its orbit revealed that it was a new planet, Uranus, the first ever discovered with a telescope. In 1801 Giuseppe Piazzi discovered Ceres, a small world between Mars and Jupiter, that was initially considered a new planet. Subsequent discoveries of thousands of other small worlds in the same region led to their eventual reclassification as asteroids.

In 1846 discrepancies in the orbit of Uranus led many to suspect a large planet must be tugging at it from farther out. Urbain Le Verrier's calculations eventually led to the discovery of Neptune.

Further discrepancies in the orbits of the planets led Percival Lowell to conclude that yet another planet, "Planet X," must still be out there. After his death, his Lowell Observatory conducted a search, which ultimately led to Clyde Tombaugh's discovery of Pluto in 1930. Pluto was, however, found to be too small to have disrupted the orbits of the outer planets, and its discovery was therefore coincidental. Like Ceres, it was initially considered a planet, but after the discovery of many other similarly sized objects in its vicinity, it was eventually reclassified as an Edgeworth-Kuiper belt object.

Observations by Spacecraft

Planets and Small Rocky Bodies

Since the start of the space age, many explorations have been performed by unmanned space missions that have been organized and executed by various space agencies. The first probe to land on another solar system body was the Soviet Union's Luna 2, which impacted the Moon in 1959. Since then, increasingly distant planets have been reached, with probes landing on Venus in 1965, Mars in 1976, the asteroid 433 Eros in 2001, and Saturn's moon Titan in 2005. Spacecraft have also made close approaches to other planets for instance, Mariner 10 passed Mercury in 1973.

The first probe to explore the outer planets was Pioneer 10, which flew by Jupiter in 1973. Pioneer 11 was the first to visit Saturn, in 1979. Following their launch in 1977, Voyagers 1 and 2 performed grand tours of the outer planets, with both probes passing Jupiter in 1979 and Saturn in 1980-1981. Voyager 2 then went on to make close approaches to Uranus in 1986 and Neptune in 1989. The Voyager probes are now far beyond Neptune's orbit, and astronomers anticipate that they will encounter the heliopause, which defines the outer edge of the solar system, in the next few years.

All planets in the solar system have been visited to varying degrees by spacecraft launched from Earth, the last being Neptune in 1989. Through these unmanned missions, humans have been able to get close-up photographs of all the planets and, in the case of landers, perform tests of the soils and atmospheres of some.

No Kuiper belt object has been visited by a manmade spacecraft. Launched on January 19, 2006, the *New Horizons* became the first manmade spacecraft designated to explore this area. This unmanned mission is scheduled to fly by Pluto in July 2015. Should it prove feasible, the mission will then be extended to observe a number of other Kuiper belt objects.

Plasma

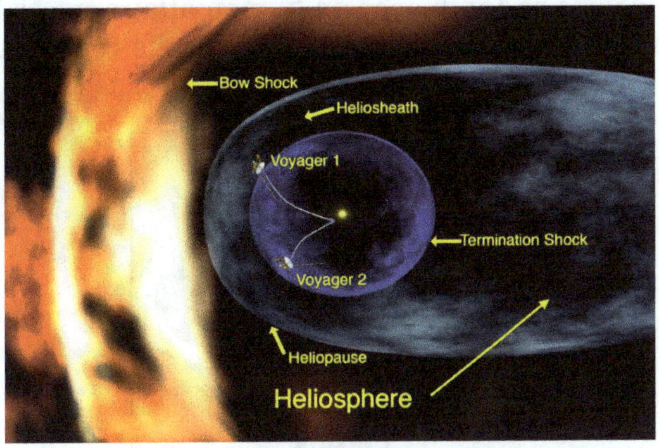

Voyagers 1 and 2 entering the heliosheath.

From the beginnings of space exploration, spacecraft have revealed magnetic fields, charged particles and electrical activity along with neutral atoms in what had been assumed to be empty space. Discovery of the Van Allen radiation belts surrounding the earth were an early clue that space was much more dynamic than could be known from the surface of the planet. Beyond these belts, spacecraft discovered a pulsing plasma stream, which was named the solar wind, and intermediate between the belts and the wind, earth's magnetopause where the influence of earth's magnetic field succumbed to the sun's magnetic field carried in the solar wind.

The story of the exploration of the solar system is incomplete without an appreciation of the extent to which exploration spacecraft are engaged in monitoring magnetic fields, charged particles, and current flows both in the vicinity of planets and in the great voids between them. In all their journeys, the exploration spacecraft have always detected the solar wind still streaming away from the sun. An important realization for life on planet earth is that the sun supports it by providing not only heat and light, but also protection from many potentially harmful cosmic rays (extremely high energy particles), which are diverted from deep penetration of the solar system by the plasma envelope of the heliosphere sustained by the solar wind.

At Home in the Solar System

To ancient peoples the sky was the home of gods and goddesses associated with what we today know as the Sun, the Moon, and the visible planets. Tracking the activities, the movements in the sky, of these deities was important for planning and conducting life activities, as evidenced by astrological systems practiced for centuries or even millenniums in places such as China, India, Mesopotamia, Egypt, Greece, Rome, and Central America. Each of these accorded central importance to the visible planets plus the Sun and the Moon, and several of these systems are still applied today. They were developed based on the assumption that the earth was the centre about which the gods and goddesses moved through the sky.

In modern times, monotheistic religions and science have together severely undermined the mystery and primacy of the deities in the sky. Conceptually, the monotheistic religions have denied their existence, while science has explained them away and replaced them with inconceivably remote and totally lifeless bodies of plasma (the Sun), rock (the Moon, Mercury, Venus, and Mars), or gas (Jupiter and Saturn). The birth of the solar system as a distinct conceptual unit stripped of its mysterious gods and influence on human affairs traces back to the radical shift some four centuries ago from the earth-centred model of the cosmos to the sun-centred solar system cosmos.

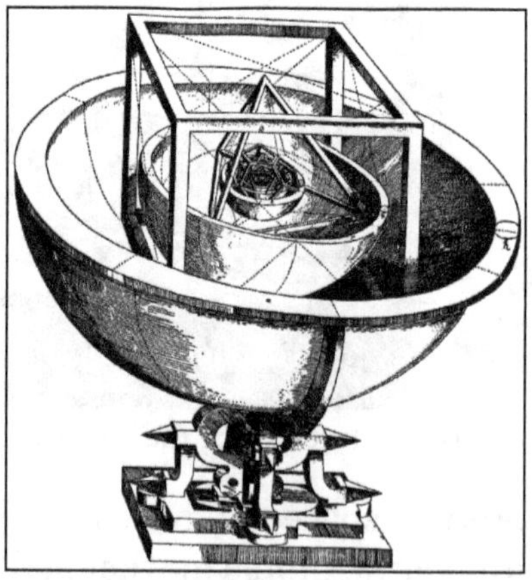

Kepler's Platonic solid model of the solar system from *Mysterium Cosmographicum* (1596).

One of the towering figures in that transition, the German astronomer Johannes Kepler, a Lutheran and still committed to the Aristotelian idea that the planets were moving on clear spheres, devised a model in which each planet moved on a sphere either circumscribed around or inscribed within one of the five Platonic solids, which were nested in an order that yielded orbital dimensions strikingly close to those determined by the astronomical methods available at the time. Kepler's proposal was one of the last serious attempts at producing a model of planetary motions that retained explicit association in this case, through the Platonic ideal forms with a higher realm of mystery and deity.

The solar system given to us by science, despite its lack of deity and mystery, has nonetheless claimed a new place of meaning in the human mind as an archetypal model of organization that can be applied beyond the realms of astronomy. This model is one of relationships in which two or more elements sharing a common purpose interact as partners, with the dominant element assuming the position of the central subject partner while the other element or elements follow the lead of (orbit around) the subject and they all together create something new. For example, the solar system, which his given birth to both life and human beings, is something new that the planets together with the sun and all the moons are co-creating.

In the early twentieth century, the first effective working model of the hydrogen atom resembled a simplified miniature solar system with one "planet," an electron, orbiting at a great distance around a massive, central "sun," a proton. Although the resemblance was forced and the model was rather quickly replaced by a quantum shell model, the concept of a tiny solar system in an atom has embedded itself in society in the icon representing an atom and atomic energy; and the solar system model is still used for elementary instruction about atoms.

Many scientists decry the continued association of the atom with the solar system model because quantum models of the atom present a nucleus that is not orbited by electrons but rather is surrounded by a nebulous cloud of quantized probable electron locations described not as an orbit but as an orbital. Disregarding or remaining ignorant of such concerns, those who view the solar system as an archetype persist in seeing the atom as embodying the same archetype when they consider that the tiny, mobile electrons and the massive, dominant nucleus collaborate in producing a new creation, the atom.

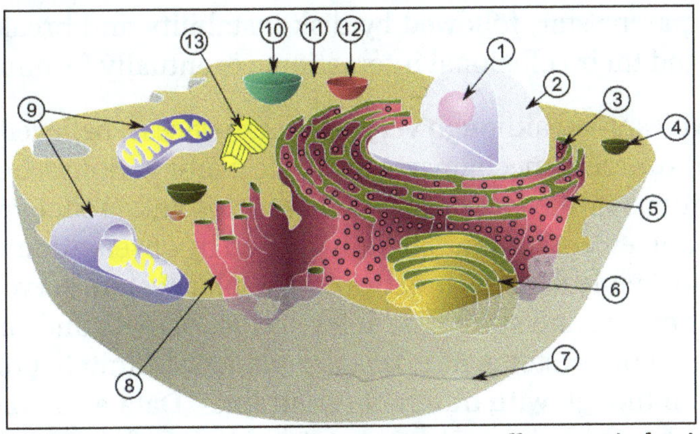

With a dominant nucleus (2) surrounded by many collaborating organelles, a typical animal cell shown here can be seen as embodying the same archetype as the solar system with its dominant Sun and orbiting planets.

A similar archetypal view can be applied as well to biological cells with their dominant nucleus orchestrating the amazingly complex metabolism carried out in the surrounding cytoplasm in a harmony of action that produces and maintains a functioning cell.

Similarly, within the human nuclear family, the parents and the children can be seen as occupying positions comparable respectively to those of the sun and the planets. Further, the parents in giving love, nurture, and guidance to the children are acting on the model of the sun gravitationally directing the earth while also providing earth with the heat and light essential for generating and sustaining life.

Viewed from yet another angle, the vertical order of the extended family generations running back from parents to grandparents and great grandparents and forward to children and grandchildren is similar to the astronomical vertical order mapped by taking the Sun to be in the position of the parents. "Generations" extending "back" from the Sun would be the centre of the Milky Way galaxy about which the Sun orbits, and beyond that the massive supercluster pulling the Milky Way toward it. Generations extending "forward" from the Sun would be the planets in the position of children and the moons of planets in the position of grandchildren. Seeing such a harmony of pattern extending from the solar system to the family, the cell, and the atom can affirm the human sense of the connectedness of all things and of the solar system as home for the human family.

Formation and Evolution of the Solar System

The formation and evolution of our solar system (and planetary systems around other stars) are among the most challenging and intriguing fields of modern science. As the product of a long history of cosmic matter evolution, this important branch of astrophysics is referred to as stellar-planetary cosmogony. Interdisciplinary by way of its content, it is based on fundamental theoretical concepts and available observational data on the processes of star formation. Modern observational data on stellar evolution, disc formation, and the discovery of extrasolar planets, as well as mechanical and cosmochemical properties of the solar system, place important constraints on the different scenarios developed, each supporting the basic cosmogony concept (as rooted in the Kant-Laplace hypothesis). Basically, the sequence of events includes fragmentation of an original interstellar molecular cloud, emergence of a primordial nebula, and accretion of a protoplanetary gas-dust disk around a parent star, followed by disk instability and break-up into primary solid bodies (planetesimals) and their collisional interactions, eventually forming a planet.

Planets formation is a widespread although very complex process, believed to be the succession of several stages affected by different mechanisms of physical interactions, chemical transformations, and numerous perturbations in the gas-dust disk. Scenarios and model approach to the origin of protoplanetary nebulae and evolution are generally backed by observational data. The mechanical, astrophysical, and cosmochemical characteristics of the solar system serve as the starting concept for the formation of planets around stars. The solar system planets and satellites architecture, as well as existing patterns in the systems of extrasolar planets definitely point to a unified process of every system formation though with different constraints. Data available on surface properties and matter composition for the solar system planets when confronting the samples of material from their embryos (small bodies) and "debris" (meteorites) provide an insight into the probable sources, paths, and chronology of this process.

Cosmochemistry and Chronology of Evolution

Of primary importance is an opportunity of getting insight into the chronology of the key physical and chemical mechanisms responsible for the early solar system evolution. Study of meteorites is the main tool of cosmochemistry aimed to reconstruct the processes of matter origin and transformations in the protoplanetary disc and forming bodies.

The time sequence was established based on the measurements of ratios of long- and short-lived isotopes and products of their decay in meteoritic materials. The main isotopic systems used in the study were U, Th-Pb, Sm-Nd, Al-Mg, Mn-Cr, Rb-Sr, I-Xe, Hf-W. Many undifferentiated meteorites (chondrites) contain the refractory inclusions of microns to cm in size enriched in refractory elements such as Al and Ca (Calcium Aluminum Inclusions or CAIs). They were assumed to belong to the ancient solid material that condensed out near the Sun ($r < 0.5$ AU) at T ~2000–1700 K. These objects, including some ultra-refractory mineral nodules, enabled a determination of the absolute age of the solar system. The measured values vary from 4567.1 ± 0.1 Ma to 4568.67 ± 0.17 Ma, with the most reliable being 4567.30 ± 0.16 Ma. Thus, the time of the solar system origin is determined with accuracy of better than ~1 Ma, or 0.02%. Concurrently, the absolute age of iron and stony meteorites of different petrological classes was defined from 1 to a few Ma younger CAI. Let us note that the submillimeter chondrules (*spherules*) embedded in stony meteorites and composed of ferromagnesian silicates are dated in the range from 4567.32 ± 0.42 to 4564.71 ± 0.30 Ma indicating an age gap between CAIs and chondrules with implication that chondrules formation lasted ~3 Ma. This time scale is in accord with protoplanetary disk lifetimes inferred from astronomical observations.

One may assume that during a few million years, interval accumulation and thermal evolution (differentiation) of the parent bodies of these ancient meteorites occurred. Provisionally the first primordial parent bodies of ~100 km in size formed in the very first few million years since the solar system origin. Such a size was sufficient for the body to experience differentiation due to intense heating by the short-lived ^{26}Al and ^{60}Fe nuclides with an iron core emergence. The subsequent core and silicate shell fragmentation caused by numerous collisions have been probably responsible for the existing iron and stony meteorites. Otherwise their existence is difficult to explain, in contrast to non-differentiated chondrites that experienced no melting by the exhausted short-lived isotopes heat source.

The above time scale is in accord with the results of computer modeling that argue that accretion of matter from the disk on the protosun terminated in 1–2.5 Ma after the system formation. The dust subdisk composed presumably of 1–10 cm particles is believed to form much earlier, in 0.01–0.1 Ma at radial distance r ~1 AU. Here critical density was achieved and gravitational instability developed. Evidently, this time was sufficient for accumulation and thermal evolution of the first solid bodies. Assuming that mass of the protoplanetary disc M_d was ~0.1 M_S and that with account for the disc partial dissipation ~0.1 M_d ultimately entered the planets, we may estimate ~10^{10} of ~100 km original bodies were born in the first ~2 Ma. This idea is in accord with the models favoring distribution of asteroids from the initial generation of planetesimals of nearly similar size on which chondrules have been presumably accreted.

Further Advancement and Current State

We shall now discuss the state of the art in our views on principal mechanisms of the solar system origin. In modern astronomy, the key consists of high resolution images and spectral features of objects relevant to planets formation at the different stages of evolution. In computer modeling, the focus is given to the theoretical treatment and development of robust models and effective algorithms allowing us to get insight into genesis of planetary system origin from primordial matter of the outer space involving disk formation, its radial/vertical compression, and dust distribution/size grow affecting the disc structure. Unfortunately, unlike protoplanetary accretion disks whose structure and evolution are accessible to astronomical observations, the mechanism of primary solid bodies set up in the gas-dust disk and their growing to planetary embryos remains rather speculative because cannot be yet tested experimentally. Hence, computer modeling is essentially the only tool to reconstruct the multiple processes involved with the use of observational constraints for models verification.

Astronomical Observations

Generally, our current understanding of the mechanism of solar system origin is in accord with the experimental data available. The discovery of circumstellar disks through high-resolution visual, infrared, and submillimeter observations, and especially extrasolar planets, extend considerably our views of the discs structure and properties of emerging planetary systems with a variety of configurations. Space telescopes such as Hubble, Spitzer, and Herschel provided an exciting picture of how all the ingredients of the cosmic stew that makes planetary systems from a protostar nebula are blended together. The concept needs, however, more rigorous and augmented support and this is the focus of the progressively refined models with an ultimate goal to reconstruct a scenario of evolution and the processes involved.

The fact that a significant fraction of young stars is surrounded by disks had already become obvious by the early 1990s, though historically the first disks were discovered around stars more massive than the Sun, such as Vega (α Lyra). The observations of young protostellar and stellar objects are currently performed over a very wide wavelength range: from X-rays to the radio band. One of the most informative methods for studying these objects is to analyze their spectral energy distributions. The studies of the infrared, submillimeter, and millimeter spectra have revealed gas-dust disks with Keplerian rotation around hundreds of T Tauri stars (young stars of less than 2 solar masses) with ages ranging from 10^5 to 10^7 yr. Gas-dust disks have been discovered around most of the observed T Tauri stars with ages 10^6 yr and around ~20–30% of the stars with ages 10^7 yr with a mean disk lifetime of 3–6 Ma. The disk masses turned ~0.01–0.2 M_0 while their extent up to ~100 AU.

Nowadays it is recognized that both young (10^6–10^7yrs) and older (10^7–10^8yrs) stars possess disks either before entering or already residing on the Main Sequence of the H-R diagram. However, young stars (of T-Tauri type) are grouped in the regions of star formation (such as Orion Nebula) where they dominate (> 70–80%) and contain more gas and dust than the older ones. Obviously, this is because the process of planet formation completed in the older disks and respectively, the major part of matter is in the solid bodies while gas was lost. Dust in these disks appears to have secondary origin as the result of numerous collisions and debris left behind after planets formation and is partially depleted by accretion on growing bodies.

Thermal infrared gas and dust emission measurements and high-angular resolution observations in submillimeter and millimeter wavelength made the most significant contribution to the study of disks' structure and chemistry, though the data interpretation is not straightforward. They revealed that the bulk of the matter is concentrated in an inner region with a radius of ~40 AU (inner boundary of Kuiper Belt Objects KBO locations in the solar system) and this similarity of parameters per se suggests that these disks with ages 10^6–10^7 yrs are probable precursors of planetary systems (Beckwith & Sargent, 1996; Cieza, Kessler-Silacci, Jaffe, Harvey, & Evans, 2005). The accretion rate (total mass flux) from the disk onto the central star to be estimated: for most stars, it lies within the range $M^{\cdot}$ ~10^{-9}–$10^{-7} M_0$ /yr with a mean value of ~10^{-8} M_0/yr; in the range of stellar ages from 10^5 to 10^7, there is a tendency for the flux to decrease from 10^{-7} to $10^{-9} M_0$ /yr (Calvet, D'Alessio, Hartmann, Wilner, Walsh, & Sitko, 2002; Hueso & Guillot, 2005). A rather low gas content in the range of 10^{-2} to 10^{-3} disk mass was found indicating low gas to dust mass ratio of the order of 10 in contrast to an earlier estimate of nearly an order of magnitude higher.

Many molecular species were identified and their isotopic compositions were measured. Numerous molecules found in protoplanetary disks (such as H_2O, CO, N_2, H_2CO, HCN, etc.) are probably genetically related to the volatiles contained in the frozen granules of primordial accreting material. They are supposed to be subsequently subjected to substantial chemical and thermal processing. Some of these molecules are apparently ionized or in non-equilibrium state due to the photolysis attributable to far ultraviolet and X-ray radiation from a young star. Observations carried out with the Spitzer Space Telescope allowed the discovery of several young stars surrounded by disks in a fairly small star-forming region. The size of some disks is comparable with the Neptune orbit in the solar system figure.

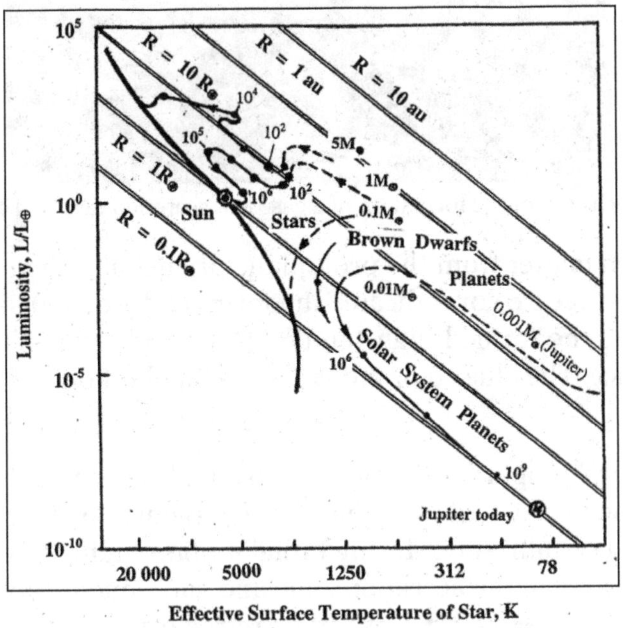

Figure: A part of the Hertzsprung-Russell diagram. Diagonal lines running from the upper left to the bottom right are disc radii from 10 AU to 0.1_0. The solid curve represents the main sequence. Tracks for protostars of the solar (larger and lower) masses are shown. Objects having $M << M_0$ do not reside in the MS.

Note that the Kuiper belt could be regarded as analogous to the low-mass dusty debris disks observed around many Main Sequence stars where existence of planets is evidenced by some disk warping.

Of great importance for understanding evolution of the protoplanetary accretion disk, in particular its inhomogeneous structure, thermal regime, and dynamics of its inner regions, are the data on dust composition and evolution. Observations were carried out in the optical, near, and thermal infrared spectral ranges and included measurements of emission spectra with the long baseline interferometry in the millimeter wavelength. Many specific features in the thermal gas-dust emissions were revealed, including sources for chemical processing of disc gaseous matter and finding spectacular dust rings. The physical characteristics and mineralogy of dust particles were inferred assuming a close analogy with meteor particles in the Earth's atmosphere. Interestingly, disk particles turned out to be much larger than micron-size dust in the diffuse interstellar medium: The largest are millimeters to centimeters across and resemble sand or even pebbles. Moreover, they exhibit the height stratification, with the smaller micron-size particles being concentrated near the disk surface. Such stratification is thought to persist for millions of years. Naturally, the content and size distribution of solid particles (granules) affects the disc medium opacity and turbulence flow patterns. They strongly influence the disk thermal regime, viscous properties, chemical transformations in a gaseous medium and, in the end, its evolution including the processes' dependence on the radial distance from the protosun and the early sub-disk formation.

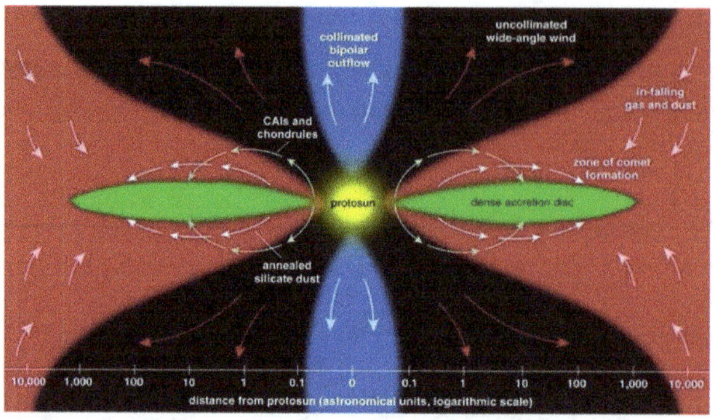

Figure: Scheme for the formation of a gas-dust accretion disk and a subdisk.

The proto-Sun onto which matter from the protoplanetary nebula (the red color) continues to accrete is at the centre. The green color indicates the forming dust subdisk, near which the outflow of gas and dust, including the formed high-temperature condensates in the inner zone, such as refractory CAIs, takes place. The blue color indicates the bipolar flows of matter attributable to the solar magnetic field.

Based on observed particles' properties, it may be reasonably assumed that their genesis is related to the formation very close to protostar and that some of them may have undergone the follow-up processes of evaporation-crystallization during radial motion in the disk. Besides, they may experience heating by shock waves in the accretion zone and subsequent rapid cooling. In accord with this concept is the finding the same type of rather refractory materials in comets, in particular those coming from the frigid outskirts of our solar system. Comets are known to have been born in regions beyond the line where water becomes frozen (called the snow line) and are referred to as remnants of the icy planetesimals. A probable scenario one may assume is that comets formed due to gravitational collapse of pebble clouds emerged through the streaming instability, which acquired the solids that would have been frozen into comets in the planet-forming disk regions.

Computer and Lab Modeling

The matter of the protoplanetary gas-dust disk is a complex system of the different phase composition, densities, temperatures, and degrees of ionization, which vary with radial distance. Basically, it is an inhomogeneous medium composed of gas and dust particles of various sizes and origin. This matter, which is generally magnetized dusty plasma, is in a state of turbulence depending on the radial and azimuthal position of a parcel of matter.When the main dynamical forces controlling the rotating disc flattening (gravitational and centrifugal) are in balance, weaker factors, such as the thermal/viscous processes, turbulence, and electromagnetic phenomena dominate the disk's evolution. They certainly affect the condensation of volatiles, including first of all water, and bear significant effect on the relative content and abundance of gaseous species and solid particles, as well as disk energetic and angular momentum transport.

When the plasma effects are disregarded, the motion of a disk medium containing dust suspended in gas can be modeled most adequately within the framework of mechanics of heterogeneous turbulized media with allowance made for the physical-chemical properties of the phases, heat and mass transport, incident radiation/opacity changes, viscosity variations, chemical reactions, phase transitions, coagulation, fragmentation, etc. Specifically, it is focused on the dynamical interaction of turbulized gas and dust including modification of the turbulence energy of the carrier phase by solid particles (i.e., the reverse effect of the dust component on the turbulent and thermal regimes of the disk gas component); the influence of turbulence on the rates of phase transitions (evaporation, condensation); on the jump-like disperse particle accumulation processes such as coagulation and fragmentation during mutual collisions between particles in the mass flow; and, finally, on the settling of solid particles through the gas to the disk midplane, where they form a flattened dust layer a geometrically thin subdisk.

Obviously, the presence of polydisperse (different particles' size) admixture in a turbulized medium complicates significantly the disk hydrodynamics, contributing to the realization of additional regimes of cosmic matter flow. Note that the synergetic collective self-organization processes in the thermodynamically open system of the protoplanetary disk against the background of a large-scale shear flow of cosmic matter associated with its differential rotation are regarded as very important mechanisms shaping the properties of a viscous accretion disk at various stages of its evolution.

Whatever is the character of events under consideration, it is clear that complex physical and chemical processes accompanying evolution of the heterogeneous medium where dust particles collisions domain, are responsible for the first solids origin and planetesimal's formation. The developed models include the sequence of changes in the aggregate state of the main protoplanetary matter components; the location of the condensation-evaporation fronts depending on the thermodynamic parameters of the disk; the role of particle sublimation and coagulation in the two-phase medium with the account for particle size distribution; the relative contribution of radiation and turbulence to the heat and mass transport; and the mechanisms for the development of streaming and gravitational instabilities with allowance made for the shear stresses in boundary layers and polydispersed, suspended dust particles.

In the most comprehensive approach, a continuum model of heterogeneous disk medium should take into account the joint influence of MHD effects and turbulence on the dynamics and heat and mass transport processes in differentially rotating matter with allowance made for the inertial properties of the polydispersed admixture of solid particles, coagulation, radiation, and changing partitioning of elements between gaseous and condensed phases. Turbulence generated at the boundaries of the protoplanetary disk layers and caused by shear flows corresponds in character to the parameters of a boundary (Ekman) layer and significantly affects the disc dynamics including the Kelvin-Helmholtz instability. It is important to emphasize that generation and maintenance of shear turbulence at various evolutionary stages of the disk involves a two-phase (gas-dust) medium with a differential angular velocity of rotation, different relative contents of dust particles, their size distribution, and coagulation processes. In general, a heterogenic mechanics approach should be applied to account for the emergence of coherent order against the background of random motions in large-scale turbulent structures. Also, the evolution of turbulence in the rotating accretion disk is supposed to be influenced by hydrodynamic helicity responsible for the cascade process of the inverse energy transfer from small to large eddies and negative viscosity appearance in the medium.

Currently, a numerical solution of the bulk of problems with allowance made for the heterogeneity of a turbulent medium, radiation, diffusion, chemistry, and MHD effects is hardly possible and only limited approaches are feasible. Note that because terrestrial planets form close to the Sun, the focus in modeling is specially narrowed to the poorly resolvable inner disk regions within several astronomical units, where matter actively accretes onto the young star. This results in the dust/gas ratio, optical opacity, and the thermal regime changes, as well as the significant contribution of photochemical processes in transformation of the matter composition and transfer.

Thermal Regime

Disc thermal regime is addressed as a key of its original inner structure and first solids formation. The viscous disc evolution with radial temperature lapse rate is regarded as the most realistic scenario of the solar system bodies' formation.

A hot rather than cold model of planets accumulation from the matter similar to chondriticmeteorites seems most plausible. It is supported by the established differences in the abundances of many elements and their isotopic ratios between the Sun, undifferentiated meteorites, and Earth. Indeed, all chondrites (except CI) and our planet are depleted in moderately volatile (Na, K, Rb, Sn, etc.) and highly volatile (Cs, Pb, etc.) elements relative to the solar abundances that are similar to their abundances in carbonaceous CI chondrites. The depletion is most pronounced in such elements as Bi, Cd, Cs, Hg, In, Pb, Se, Te, Tl, Zn, S, etc. that was found to be typical not only for various types of chondrites but also for the bulk composition of the terrestrial planets and some large planetesimals, for example, for the parent bodies of basaltic achondrites eucrites. Hence one may conclude that the differentiation of moderately and highly volatile elements was an important large-scale process at the early evolutionary stages of the protoplanetary accreting disk under relatively high temperature.

Basically, the observed depletion of highly and moderately volatile elements could result from either partial evaporation or incomplete condensation of the proto-matter of planets and the parent bodies of chondrites, because the higher the volatility of an element, the greater the depletion. Thermodynamic calculations showed that temperatures of no lower than 1200–900 K are required in both cases. Under such temperatures only non-equilibrium partial evaporation may occur and indeed, some CAIs show clear chemical and isotopic evidence of that. It is of interest to note that an experimental study of the fractional evaporation of the material of CI chondrites aimed at obtaining the material of chondrites of other types showed that, irrespective of the redox evaporation conditions, the residue obtained through heating differs radically from the actual chondrites material in the abundances of highly and moderately volatile elements. These differences were claimed to be particularly true for the abundances of such pairs of elements as Zn and Se; Sn and Pb; Rb and Cs; etc., with similar degrees of depletion relative to the abundances in chondrites. The results, however, are not unambiguous because all chondritic meteorites are cosmic sediments; that is, their components have formed at very different conditions within the nebula and experienced different thermal history before being accreted together later.

The condensation mechanism for the differentiation of moderately and highly volatile elements accompanied by their incomplete accumulation in the proto-matter of planets and the parent bodies of chondrites seems more justified. For most elements, there is a clear correlation between the degree of depletion and the temperature of its condensation from a solar-composition gas. It is most likely attributable to the influence of the kinetic constraints on the heterogeneous reactions in a gas-solid system associated with the reduction of the reaction surface of small dust particles during their accumulation and some other factors. This allowed us to conclude that much of the chondrites material formed through the condensation of the gas phase of the protoplanetary disk.

Temperature initially increased at the disk formation stage, reaching the evaporation temperature of magnesian silicates and metallic iron at a distance of 1 AU (1400 K) and then decreased in the course of viscous disk evolution. In the formation region of the terrestrial planets, it has always remained well above the H_2O condensation temperature it did not dropped below 300–500 K.

Apart from mineral phases containing moderately volatile elements, there are crystalline Mg- and Fe-bearing silicates in chondrites that support the hot protoplanetary disk formation model. One may assume that these amorphous and crystalline silicates formed as a result of the condensation processes in the inner zone of the circumstellar disks. Amorphous silicates of the interstellar molecular cloud from the fragment of which the solar system formed are believed to have evaporated in its inner zone at the early evolutionary stages to be condensed as the nano-to-micrometer sized crystalline silicate grains during subsequent cooling of the disk gaseous phase. One may suggest that the inclusions enriched in the most refractory elements (such as Ca and Al composing CAIs, as well as the rare Hf, Sc, Lu, etc.) were the earliest condensates formed near the Sun (r < 0.5 AU) at T ~2000–1700 K and partially carried outward through the radial drift up to the formation zone of the parent bodies of chondrites with refractory chondrules embedded in their matrix and entered rocky planetesimals. Obviously, a small fraction of them reached r ~5–10 AU, as evidenced by the findings of olivine crystals in the material of Halley and Hale-Bopp comets as well as in the Comet 9 P/Tempel nucleus investigated in the Deep Impact experiment.

Several important formation stages of the gas-dust disk around the proto-Sun and its subsequent dynamical, thermal, and cosmochemical evolution, including the non-equilibrium partial evaporation, condensation, and compaction of protoplanetary matter were traced in the framework of quite comprehensive self-consistent numerical model. In particular, it yielded the distributions of thermodynamic parameters (T P) in the viscous disk around the young Sun passing through the *T Tauri* stage, as a function of radial-vertical distance (the rz coordinates). Radial temperature profiles in the disk's mid-plane at the different stages of its evolution were derived, as it is shown in figure.

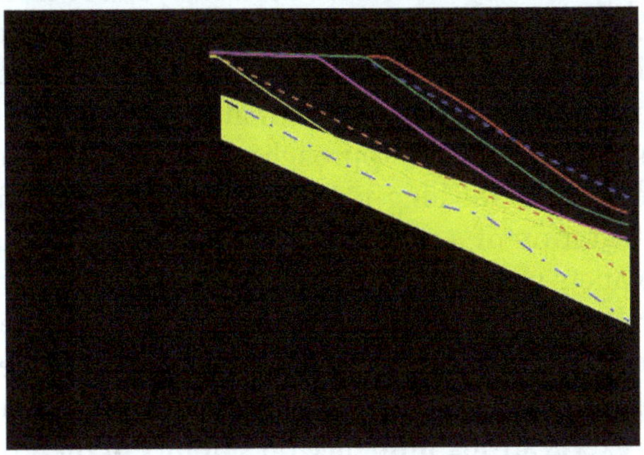

Radial temperature distribution T*(r)* in the midplane of a protoplanetary disk between 10 and 0.5 AU. 1–4 formation stage between 0.12 and 0.45 Ma (measured in millions of years) from the disk origin; 5, 6 viscous dissipation stage between 0.65 и 1.8 Ma; 7a–7в stages of dust subdisk compaction and dust clusters formation at about 2 Ma.

Temperatures are controlled by the processes of condensation/evaporation of forsterite, enstatite, and iron; maximum values Tsi ≥ 1,600 K at r = 0,1 AU and Tsi ≤ 1,400 K at r = 1,4 AU. Water ice is formed at Tw ~140–160 K under gas pressure P = 10^{-7}–10^{-4} bar at about 3 AU (break in the dot-dashed curve). The green zone shows the range of possible temperatures at the stage of dust clusters formation.

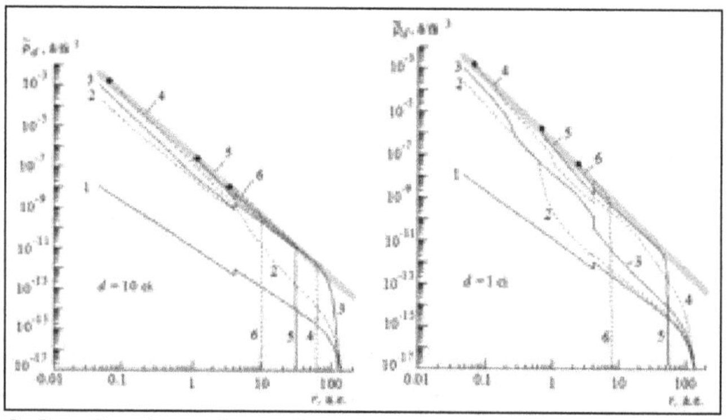

Figure: Variation of subdisk dust density ρ_d depending on the radial distance r from 100 AU for particles of mean size 10 cm (left) and 1 cm (right). Broad diagonal band critical dust density ρ_{dcr} corresponding to the disk gravity instability and its fragmentation into dust clusters. Curves 1–6 – time sequence 0, 1×10^3, 5×10^3, 2×10^4, 5×10^4 and 1×10^5 (yr) from the beginning of subdisk evolution.

Variation of subdisk dust surface density ρ_d as a function of the radial distance r, the dust/gas densities ratio and size of dust particles is shown in figure as a sequence of subdisk dust density evolution, before critical surface density is achieved and the subdisc break up into dust clusters.

Dynamics

It should be specially noticed that solar system formation was intrinsically related to the dynamical processes at all stages of origin and evolution including different kind of resonances defining the system's structure, configuration, and stability. Complex nonlinear dynamical processes impose certain constraints on the models of disc evolution and planetary embryos origin. The role of dynamics is manifested, in particular, by observations of the formation of some peculiar configurations in the extrasolar planetary systems around the parent solar-type stars.

The basic principles of the dynamical systems were derived by Anri Puankare and thoroughly developed by the numerous researchers. The main parameters defining planetary system configuration are resonances and dynamical chaos. The study is used to be performed as general N-body problem or in the framework of restricted three-body approximation with involvement of some constraints. Besides resonant and chaotic motions, this approach allowed us to describe periodic orbits, collisions, escape, and other processes of the long-term dynamics based on either analytical or (mostly) on the direct numerical integration of a set of ordinary differential equations. It was shown that orbits of the solar system planets with small eccentricities and inclinations are weakly chaotic and do not exhibit essential secular variations on the time scales comparable with the solar system age. Resonances (a near-exact relation between characteristic frequencies, e.g., orbital periods, of two bodies) and tidal oscillations are pronounced in the systems of small bodies and their tidal interactions with planets, which exerts strong influence on stability of their orbits being responsible for migration processes. The stability problems including interaction of nonlinear resonances in Hamiltonian systems were analyzed in detail in the framework of Kolmogorov-Arnold-Moser (*KAM*-theory). Recently, these and some other studies laid the basis for investigation of the structure and dynamics of the systems of exoplanets around binary and multiple stars.

Migration involving both forming planets and small bodies could play a major role in the history of early solar system affecting its configuration and planets composition. Great diversity of bodies revealed by compositional mapping of the main asteroid belt implies substantial mixing of bodies of different taxonomic classes through the various dynamical processes affected by planetary migration. The outer planets' instability could cause massive influx of bodies to the inner solar system. Hydrated C-type asteroids and comets transported from outer regions of the solar system would contribute to the above diversity and also be responsible for delivery of water and volatiles to the Earth and terrestrial planets deposited as a veneer at the final stage of these planets' accumulation.

Unlike classical models of the terrestrial planets evolution with Jupiter–Saturn's stable orbit positions, there were different scenarios developed in order to explain inner solar system's water under action of the outer planets' dynamical instability. Migration of bodies and dust from the disturbed trans-Neptunian regions in Jupiter Crossing Orbits (JCO) and then in the main asteroid belt to eventually become NEOs was studied in the numerical models. The mechanism involving unstable positions of the giant planets in the course of their formation was explored by several teams. Inward migration of Jupiter from ~ 20 to ~ 5AU and getting it captured in 2:3 or 1:2 resonance with growing Saturn (which could halt and reverse the inward migration of Jupiter) was considered in the model known as the Grand Tack scenario. It was assumed that this process would shepherd outer belts' bodies inward and exert influence on planetary embryos in the inner region. This would result in volatile implantation through planetesimals' impacts during terrestrial planets formation and storage of remnant pristine bodies in the main asteroid belt, as well as the position and coexistence of asteroids of different compositions. Moreover, such a scenario was invoked to explain depletion of water in the zone of terrestrial planets' formation in terms of a "fossilized" snow line position outside Jupiter's precursor orbit beyond the snow line at ~3 AU preventing condensation of water and ice inside and/or mechanism of water delivery related to giant impacts.

Protoplanets embedded in a gaseous disc would experience drift changing their orbital positions. The movement of Uranus and Neptune from the place of their initial formation close to the Jupiter–Saturn zone further away from the Sun was assumed to exert formation of the trans-Neptunian Kuiper belt. This scenario was justified by evaluating the time required for the accretion of Uranus and Neptune in their present-day orbits. Indeed, modeling showed that such a time would exceed the age of the solar system. The model, which is sometimes referred to as reconfiguration of giant planet orbits suggested an existence of the primordial disk of several tens of Earth masses made of comet-like objects residing just outside the initial orbits of the giant planets. The disk is thought to be scattered throughout the solar system by gravitational interactions between the giant planets leading also to their migration. It is worth noting that movement of Uranus–Neptune outward due to gravitational interactions with the planetesimals would ultimately eject them into hyperbolic orbits.

The dynamical models involving instability in the giant planets zone were associated with the dramatic event of Late Heavy Bombardment (LHB) of the Moon and terrestrial planets and the Moon crater chronology. However, timing of the instability (well before the LHB or later) underlying these models remain unclear. Most recently, instead of using the terminal cataclysm hypothesis that poorly explained the triggering flux of bodies from the outer asteroid belt, linked the source of LHB projectiles with leftovers of planetesimals ("tail end of planet accretion") from the time of outer planets' formation. In support of this idea, they brought some evidence of lunar craters' records and matching density of craters with the Highly Siderophile Elements (HSE) in the lunar mantle. These "iron loving" elements were found in chondrite meteorites and presumably they were delivered with projectiles of similar composition after a major event of the lunar matter differentiation, with the core emergence and sinking of the original HSE in the core. Hence, geochemical constraints of large craters' composition formed by projectiles and the timescale of mantle ocean crystallization (~ 4.35 Ma, with the account for Fe-S segregation), advocate for the imprint of LHB event at the tail end of planet accretion. The modern model coupling LHB at ~ 3.9 Ma with the time of outer planets' instability appears to accommodate the current knowledge with minimum assumptions.

The critical role of C-chondrites and water transport in the inner solar system could be played by the process of outer planets' formation with no change of their orbital positions. This is a kind of scattering model that postulates a slow accretion of gas in the broad outer belt rich of volatiles (~5–20 AU) by the giant's cores until some threshold with the onset of Jupiter's runaway growth. This process provides destabilization, gravitational scattering of nearby planetesimals by Jupiter and Saturn and their large-scale radial mixing, some of them moving inward and residing (resonance) orbits' interior to that of Jupiter. As gas dissipates, orbits of the disc planetesimals become progressively eccentric, including those trapped in the main asteroid belt and crossing the region of growing Earth and terrestrial planets, thus delivering water with C-chondrites and other type of meteorites. Later implantation of these species by scattering triggered by growth and possible migration of Saturn and icy planets turns out to be non-efficient. This scenario with no or very limited key assumptions related to the Grand Tack or Nice models seems quite plausible and implies a close relationship of both inner and outer planets' formation.

Particles Growth and Solid Bodies Formation

The key problem of the solar system origin is how the solar system bodies (original dust and condensates, as well as those produced in coagulation) progressively grew on scales ranging from nano- and micron-size particles to planetesimals and planets thus ranging over dozens orders of magnitude in mass. As mentioned above, time-dependent modeling with the different particle size distribution functions and limited lab experiments is the only tool to gain insight into the problem. Numerous attempts to reconstruct the process taking into account nebula thermal structure; evaporation fronts (EFs') position for different components depending on radial temperature lapse rates in the disc's midplane; particle growth by sticking limited by bouncing; fragmentation and radial drift due to nebula headwind drag of growing bodies influencing mass redistribution; etc., have been undertaken. Also, lab experiments with silicate and ice particles collisions in microgravity conditions revealed some important patterns of particles and particle aggregates formation. Some important constraints were deduced for particles bouncing and sticking, including translational energy and coefficients of restitution estimates depending on impact parameters.

According to the modern views, the process started owing to the above-mentioned streaming/gravitational instability development in the dense subdisk formed out of the dust component settled down to the midplane of the gas-dust disk. This was followed by the primary porous dust clusters formation from which the first solid bodies of pebble-boulder size and eventually planetesimals of asteroid size have emerged. Some of these cm-size and larger pebbles were possibly assembled into porous clumps giving birth to comets; this approach accommodates the primordial rubble pile theory though other theory postulates that comets were made out of debris left over from the main planet-building phase, thus preserving remnants of the protosolar nebula matter.

Nonetheless, collisional rubble pile theory better explains a rather small size of comet's nucleus (~5–10 km) and even the bi-lobed shape of some of them, because of their gradual formation at low speed of leftover pebbles/grains after larger bodies have accumulated and gas in disc has disappeared, followed by gentle collisions owing to stirring the cometary orbits, specifically at the skirt of planetary system.

The sequence of solids growth and timescale of such a scenario is shown schematically in figure below.

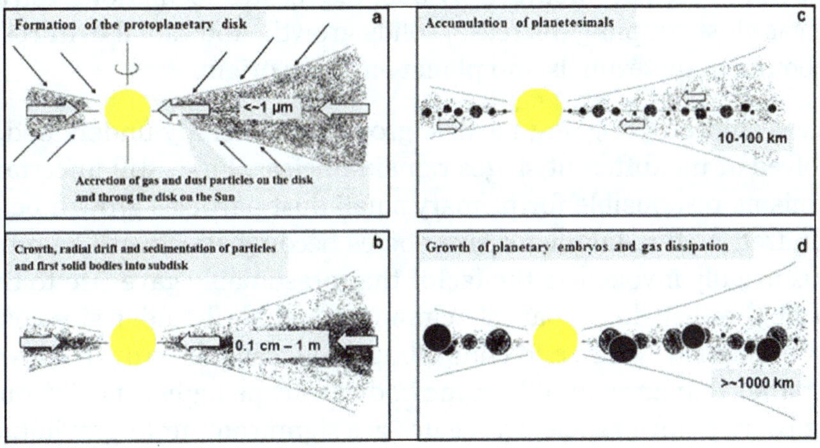

Sequence of the of protosolar disk evolution.

(a) Disk formation due to accretion gas and dust from the protosolar nebula and protosun emergence in the centre. (b) Disk flatteniand dust particle sedimentation toward the midplane and dense dust subdisc formation; particle growth. (c) Subdisc fragmentation into dust clusters due to streaming and gravity instability development, cluster collisional interaction and solid growth, including first proto-planetesimal accumulation with gravity domain. (d) Planetesimals and planetary embryo formation, gas dissipation, and original solar system architecture setup evolving ultimately to the contemporary configuration. Time span: (a)–(b) 10^5–10^6 yr; (c)–(d) 10^6–10^7 yr.

The initial process is thought to last less than 10^5–10^6 years of the overall ~10^8 years of planets formation. Note that the same short timescale is assumed for comets' formation as well as for initial growth phase of the TNOs probably influencing the cometary-like bodies origin and evolution. The shear turbulence mechanism would promote ring-like contraction of dust in protoplanetary cloud into a thin disc ($h << r$) of non-regular shape widening toward the periphery. Solid bodies becoming planet embryos are formed from initially "loose" gas-dust (porous) clumps filling the main part of their sphere of attraction (Hill's sphere) and slowly contracting due to internal gravitational forces. Let us recall that whereas the growth of particles during their collisions is hampered in chaotic turbulence, their coalescence and enlargement can occur inside turbulent eddies' coherent structures promoting dust clusters set up in vortexes of baroclinic nature in a broad range of Stokes (St) number defined by the ratio of particle drag in an ambient gas to characteristic time of the system (defined as inverse Kepler frequency Ω^{-1}) and dependent on size and density of a particle. Solid particles may be also concentrated in the disks with quite weak turbulence where denser than average matter structures (with large dust-to-gas ratio) would create a large population of aggregates and trigger streaming instability. This could be the case in outer zone of massive discs where rapid growth of aggregates to planetesimals was suggested.

Generally, both instability mechanisms support the basic Safronov and Goldreich-Ward ideas about disc viscous accretion and subdisk matter gravitational collapse. As we have seen, in the modern models, the focus is also given to the competing processes of gas and dust photo-evaporation by the solar EUV-X-ray radiation and gas condensation/condesate growth, as well as to dust trapping, clustering, and coagulation including particles size distribution and stratified turbulence. These processes are accompanied by dust aggregates formation/restructuring through particles sticking, gas-dust coupling decrease, solids growth, and radial drift resulting ultimately in gravitationally bound planetesimals and planet embryos origin.

Although the basic scenario of ongoing particle growth is generally understood, many details of the processes involved at the different stages remain unclear. First, still uncertain are the details of physical mechanisms responsible for primary small dust particles growth before gravitational interactions of hundred meters-kilometer size bodies become effective. The process of particles' mutual collisions is usually invoked as the factor that presumably gave rise to the aggregation of small particles to yield either dense particle clumps or pebble/boulder-size compact aggregates migrating in the protoplanetary disc and controlling planetesimals' growth. However, a probability of destruction rather than growth of forming bodies may be higher if collisional relative velocities and bouncing barriers for particle aggregates are significant, until gravitational attraction of planetesimals becomes dominant. Hence, an efficiency of sticking of dust particles into aggregates through collisions is a rather delicate mechanism, specifically if equal size particles of ~mm size are considered. The formation of larger objects limited by bouncing is thought to be augmented by the streaming instability, gravitational collapse, collective particle behavior, and/or by static compression of fluffy dust aggregates.

As a compromise ensuring higher efficiency of agglomerates creation we advocate for collisional integration of fluffy clusters made up by micron-sized particles rather than individual particles themselves, in support of the lab experiments mentioned above. It is further assumed in the developed models that clusters and their individual particles have fractal structure. Such an approach was thoroughly explored and validated theoretically involving concurrence of gravitational and brownian coagulation of dust monomers, aggregates growth, and porous bodies interaction/growth.

The bottom line of our numerical model is the agglomeration of small particles in the primary fluffy dust aggregates of low-bulk density and fractal nature of the latter, giving rise to their compression and formation of first dense bodies of progressively larger size. Unlike an approach used in the previous simulations, the mode of particle on-head and offset collisions inside dust clusters was more strictly assessed in numerical N-body models. Method of permeable particles and modified Newton model of collisions were utilized in terms of restitution coefficient Kr dependent on the distance between centres of particles and their relative velocity, taking into account internal structure of particles in fluffy clusters and complicated patterns of their interactions, specifically in the contact zone. This enabled us to examine collisional evolution of fluffy clusters beginning from nano-particles sticking together through electrostatic forces and growing up to compressed aggregates of larger size and different patterns. Example of 3D modeling of evolution of fractal dusty clusters with different numbers of populated submicron particles is shown in figures.

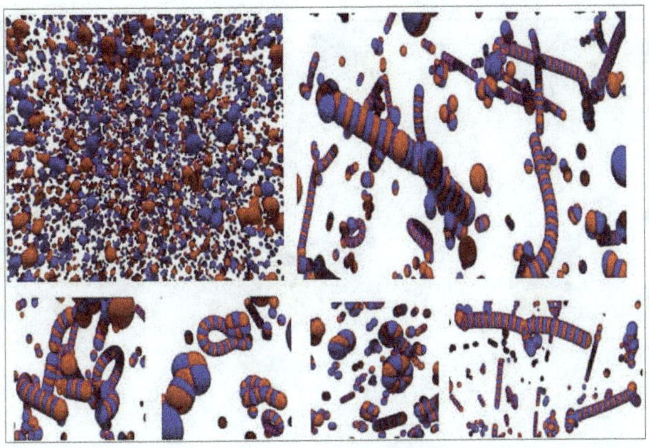

*Figure:*3D-numerical modeling of submicron particles, collisional interaction, and growing within a grid in the primordial dusty cloud (density $d = 1.5 \times 10^{-21}$ g/nm³. (a) An example of the original 3D structure of a selected grid. (b) Examples of agglomerate formation in the process of collisional evolution of particles in some grids.

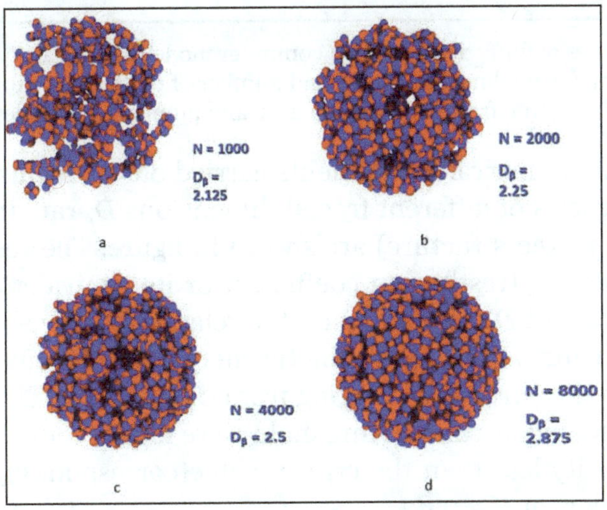

Figure: 3D structure of fractal dusty clusters with different numbers of populated particles evolution (a–d) Red and blue colors denote positively and negatively charged particles of about 20-nm characteristic size in a quasi-neutral medium.

Figure is a snapshot of a cross-section of randomly selected grids for fractal dusty clusters with different numbers of populated particles in figure.

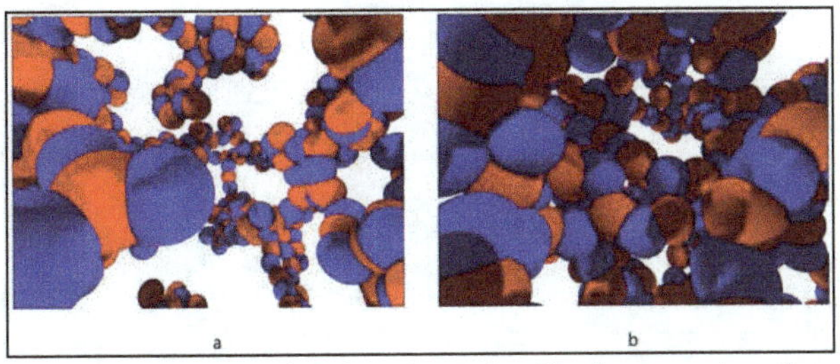

Figure: Snapshot of a cross section of a randomly selected grid when getting through fractal dusty clusters with different numbers of populated particles (clusters a and b, respectively).

Of special interest is the evolution of fluffy dust clusters in mutual collisions that is critical for dusty agglomerates formation, as shown in figure.

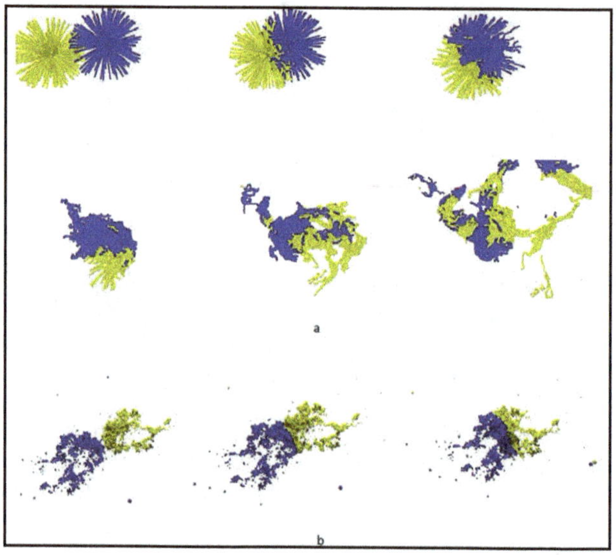

Figure: Evolution of the structures of fluffy dust clusters; computer modeling of collisional interactions. (a) Fluffy clusters of characteristic size 50 nm, fractal dimension 2.55, and number of particles = 8,192. (b) Fluffy clusters of characteristic size 75 nm, fractal dimension 2.15, and number of particles = 3,072.

Examples of the respective numerical experiments carried out with clusters composed of various number of submicron particles of different fractal dimensions D_β ranging from 2.025 (very fluffy structure) to 2.975 (well-packed structure) are shown in figure. The results of modeling of cluster interactions reveals how K_r (restitution coefficient or its equivalent collision recovery ratio) changes for clusters of different D_β depending on the relative distance between colliding *i-j* particles (0–1), and collision energy E_{ij}/E_{int} (E_{ij} is kinetic energy of collision; E_{int} the internal energy of molecular coupling preserving structure) ranging from 0.01 to 0.99. The strong dependence of K_r on the parameters involved and are bound threshold were found, for example, for dense colliding clusters with relative velocity less than the critical value(corresponding to the energy of plastic deformation)bouncing occurs at $K_r > 0.87$.

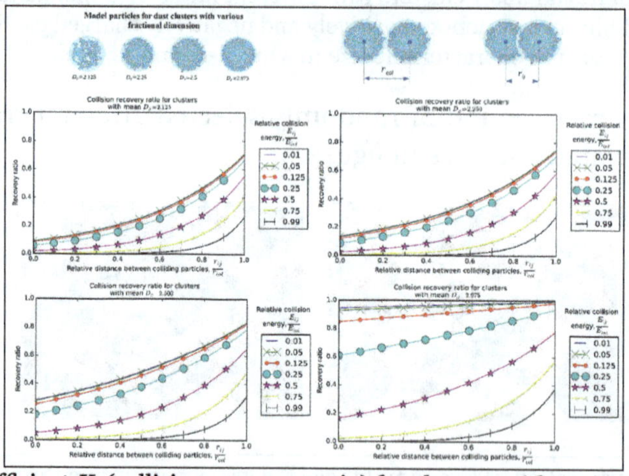

Figure: Restitution coefficient K_r (collision recovery ratio) for clusters with various fractal dimensions $D\beta$ as shown here. Note: r_i, r_j — radii of *i* and *j* model particles; r_{ij} — distance between *i* and *j* colliding model particles; $r_{col} = r_i + r_j$ — collision distance; $\frac{r_{ij}}{r_{col}}$ — relative distance between *i* and *j* colliding particles; *E int*: cluster's internal energy; *E ij*: kinetic energy of collision; $\frac{E_{ij}}{E_{int}}$ — relative collision energy.

The results are regarded as an important milestone toward in-depth study of "cluster-cluster coagulation" at the initial stage of the protoplanetary gas-dust disc evolution and results are addressed as precursors of primary objects growing. They are in accord with the conclusion that a cluster with a larger number of particles is harder to destroy (easier to survive) in energetic collisions. Nonetheless, the problem is still far from being solved, especially if one keeps in mind that the global qualitative effect of disk gravity further increases collision/impact velocities and adds additional jitter to the orbital evolution of the primary bodies.

The next stage of growth of primordial seeds of bodies to planetesimals, with the allowance for mechanism of self-gravitation, seems more conceivable. At the first approximation, the mass distribution of particle agglomerates and proto-planetesimals obeys the known Smoluhovsky coagulation equation (a sort of the Boltzmann kinetic equation when dealing with coagulation process) with the account for gravity mutual attraction and fragmentation in non-gentle collisions. This interaction along with the dust sedimentation onto the formed bodies residing on intersected orbits (with chaotic velocities superimposed on quasi-circular orbital velocities) results in further growth of these protoplanetary embryos followed by the gradual scooping up of smaller bodies in due course of the swarm evolution. It is supposed, however, that there is a sort of barrier at around one meter size-range because of the effects of body destruction and their inward drift toward the central star. The latter is caused by aerodynamic braking of these bodies in the remaining gas whose rotation velocity becomes lower than the Keplerian one. Primordial bodies beyond the meter-size barrier to proto-planetesimals growth through "coagulation" after more complete gas evacuation seems more feasible.

Obviously, competitive processes of destruction and particle trapping during radial matter exchange, along with particle concentration via streaming and/or gravitational instability, eventually results in bodies' growth. Based on the results of modeling supported by lab simulation millimeter- to centimeter-size pebbles in a gravitationally bound collapsed cloud (due to, for example, disk streaming instability) would experience numerous collisions with different outcomes depending on their relative speeds and the cloud density, resulting eventually in a quite intense formation of planetesimals. As far as a planet's accumulation is concerned, gravitationally focused mutual collisions of asteroids and planetary embryos are thought to be responsible for the emergence of progressively growing proto-planets possessing larger gravitational potentials (oligarchs), such an oligarchic growth being accompanied by catastrophic fragmentation of other bodies in a swarm of planetesimals. A system of satellites around a proto-planet could be explained invoking the same mechanism.

A possibility of the formation of bodies from dust clusters in mass range from 10^{20} to 10^{22} g (corresponding to asteroids of tens to hundred kilometers in size) was shown in the numerical model. A mechanism of gravitational collapse of massive swarms of small particles and their pile-up without passing through the intermediate phases of growing was proposed to explain early formation of the ~100 km bodies, a source of very old iron meteorites, thereby circumventing the meter-size barrier problem. Similarly, a prompt planetesimal's formation in regions beyond the snow line due to streaming instability-induced gravity collapse of radially drifting millimeter-size dust particles to ensure material deposition for the giant planet cores was suggested. Anyway, whatever particular solution of the problem, further growth occurs through interaction of planetesimals with solid and gaseous components under gravity domain and sporadic mutual collisions.

Note that in the numerical experiments, some constraints were placed on the value of angular momentum of the porous (low density) gas-dust clusters (pre-planetesimals) of about Hill sphere in size and their moving before collision along heliocentric orbits. The model aimed to assist dynamical formation of planetary/satellite bodies with application to trans-Neptunian system. Arguments were drawn for consistency of the properties of wide binaries in the Kuiper Belt with a primordial origin during gravitational collapse. As the outcome, origin of the Earth-Moon system was attempted to explain in terms of the compression of a low-density cluster of 0.1 mass of the contemporary Earth formed from the collision of two original clusters on the close heliocentric orbits, which allowed the system to acquire the necessary angular momentum.

As it was said above, significant role at the final stage of planet's growth could play migration and resonances in the forming planetary system. The commonly adopted scenario of the solar system origin involving planetesimals growing to planet embryos together with asteroid-size and comet-like bodies as remnants of planets is illustrated in figure.

Artist's concept of a planetary system formation, as seen from the outer edge of the gas-dust disc around an emerged central protostar. Planetesimals growing to planet embryos, together with asteroid-size and cometlike bodies as remnants of planets' formation, are seen. The picture resembles the commonly adopted scenario of the solar system origins.

It reflects our current general understanding of how original bodies and then planets formed though we are still far from resolving this mysterious subject.

Inner Planets Evolution

The key question intrinsically related to the origin of the terrestrial planets is why the three neighbour planets Earth, Venus, and Mars inheriting disc composition in a close neighbourhood and basically, exhibiting similar chemistry in the formation took different evolutionary paths? Namely, what caused, unlike the pleasant climate of Earth, hostile thermal regime of Venus, and why the assumed original clement climate on Mars has drastically changed?

To answer this question, many complicated problems closely related to planetary cosmogony should be addressed. Some of them relevant to this important topic were considered in previous chapters of the paper and they are discussed in more detail in some reviews and monographs. Here, we focus only on limited aspects of the problems involved that concerns a rather simple approach in the treatment of thermal regime formation on the inner planets and the fate of water as the key climatic factor impacting on their evolution.

Let us first note that Earth is located in a comparatively narrow zone of circumsolar space where the development of favourable (for life's existence) climatic conditions is possible. This is the so-called habitable zone in the solar system whose inner boundary is located 10–15 million km closer to the Sun than the Earth's orbit while its outer boundary extends to approximately the orbit of Mars. The orbit of Venus turns out to be outside this zone, at a distance that is almost triple the critical value. Obviously, if the Earth were moved to the place of Venus (actually, even less) then it would probably evolve according to the Venusians' scenario.

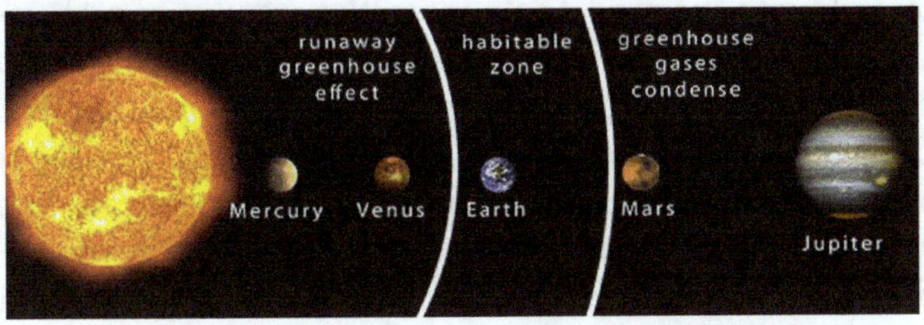

Figure: The habitable zone in the vicinity of Earth.

The zone is limited by only about 10–15 million km toward the Sun and extends outward to nearly the Mars orbit. The planet shifting closer to the Sun would meet runaway greenhouse similar to that on Venus (T_s = 735 K; P_s = 92 atm), whereas at Mars's position, greenhouse gases would condense out and the atmosphere would collapse. Unlike Earth, the atmospheres of Venus and Mars consist mainly of carbon dioxide; their abundance is controlled by the carbonate-vollastonite equilibrium between atmosphere and crust, which is strongly dependent on surface temperature T_s. CO_2 storage in the Venus atmosphere is the same as what is locked in the Earth's crust, and it would release into the atmosphere amounting to P_s ~ 90 atm if the clement climate on the home planet irreversibly changed (temperature dramatically raised) and the CO_2 balance maintenance ceased.

Based on the average annual balance of the solar radiation absorbed by the terrestrial surface and atmosphere with an account of its albedo the effective (equilibrium) temperature of Earth is T_e~255 K, which is even below the freezing point of salt water. This means that favourable climate conditions could hardly have arose on the home planet unless the greenhouse effect caused by absorption of solar radiation in its relatively thick atmosphere would raise the mean temperature near the surface to the pleasant 288 K. This atmosphere of secondary origin, unlike the primary atmosphere captured from the protosolar nebula and composed of light atmophile elements that were lost, formed owing to degassing of volatiles from the interior and from collisions with remnant planetesimals and migrating bodies. It presumably contained carbon dioxide, nitrogen, water vapor, and some minor species including such greenhouse gas like methane.

Earth could therefore retain its water on the surface, which was mainly concentrated in the oceans. In turn, carbon dioxide would accumulate incarbonates of sedimentary rocks through its binding with metal oxides incorporated into the ocean crust and upper mantle (with the formation of aqueous silicates) and biogenically, through the deposits of lime skeletons of sea organisms. Dramatic change of the Earth's atmosphere with its modern composition occurred much later, about 2.8 Ma, due to bacterial production of oxygen accumulated in the planet's gas envelope, which made possible the life forms rich diversity and proliferation.

In contrast, equilibrium temperature of the early Venus is supposed to be not lower than T_e~325 K, which is above the water boiling temperature in a low pressure environment. One may reasonably expect surface pressure on ancient Venus as an Earth-like plane to be substantially lower than 1 bar. Under these conditions, carbon dioxide would gradually accumulate in the atmosphere together with water vapors, which in turn, have contributed to a further rise in surface temperature. The runaway greenhouse eventually developed resulting in the transport of increasingly large amounts of CO_2 and H_2O into the atmosphere, up to some equilibrium state corresponding temperature 735 K and pressure 92 atm. At such a temperature, carbon dioxide turned out to be not bound in carbonates of sedimentary rocks, as on Earth but was released into the atmosphere giving rise to very high pressure. According to the model estimates, the amount of carbon dioxide locked in the Earth's sedimentary rocks is comparable to the content of CO_2 in the Venus atmosphere and it would constitute nearly the same pressure had the Earth's temperature raised to similar temperature and CO_2 released in the atmosphere. The latter is characterized by the relationships between mineral phases and volatiles on the surface, with the carbonate-silicate interaction in the upper layer of the planetary crust being the most important. The above scenario is characteristic of a system with positive feedback and internal instability when initial perturbation is not suppressed but, on the contrary, fairly rapidly enhanced.

As far as water is concerned, its accumulation on Venus with the involvement of both endogenic and exogenic sources, likewise on Earth and Mars, would result in the initial amounts of water nearly equal to the volume of the terrestrial hydrosphere. However, water could not be preserved on the Venus surface at the temperature above the critical point (647 K), even as salty aqueous solutions (brines) with a slightly higher critical temperature (675–700 K). As to the atmosphere, the amount of water in it is negligible. Thus, the key problems of the planet evolution are whether water was on early Venus, where it was stored, and, if so, when and how was it lost. The deuterium-to-hydrogen ratio measured in the Venusian atmosphere, which turned out to be larger than that in the Earth's atmosphere by two orders of magnitude, confirms the concept of ancient Venus ocean because high deuterium enrichment can be explained by an efficient thermal escape of the lighter hydrogen isotope and leaving behind heavier deuterium. Note that adoption of this mechanism requires an assumption that enormous masses of liberated hydrogen and oxygen through dissociation by the solar EUV radiation were removed from the atmosphere by efficient dissipation and bound by surface rocks, respectively, which seems, however, hardly realistic.

There is also a different viewpoint that Venus was initially formed as a "dry" planet, though the introduction of volatiles through the mechanism of heterogeneous accretion is assumed to play an important role as well. In this scenario, the water content was essentially constant throughout the geological history of Venus, remaining approximately at the current level while the efficiencies of heterogeneous accretion and hydrogen dissipation from the atmosphere were considerably lower. In such a case, the observed enrichment of the Venusian atmosphere in deuterium during the isotope fractionation in the process of their dissipation from the atmosphere can be also explained.

Addressing now the problem of Mars evolution, we start from a well-supported concept that the quite pleasant natural conditions presumably were on Mars at early epoch, which is backed by the morphological features preserved on the planet's surface as well as by some other evidence. Since equilibrium temperature of early Mars would not exceed T_e~220 K it would be able to retain only frozen water on its surface. However, there is widespread evidence of the presence of liquid water flows meaning considerably denser atmosphere and much more favourable climate ~3.8–3.5 billion years ago. Obviously a catastrophic drought suddenly occurred on Mars rather than a long-term evolution has resulted in the present-day natural conditions. This dramatic change cannot be explained by the fact that the planet is 0.5 AU farther away from the Sun than Earth. More likely it should be associated with the fact that the mass of Mars is almost an order of magnitude lower than that of the Earth implying a lower storage and an early depletion of the former in the long-lived radiogenic isotopes, which served as the main energy sources responsible for the thermal history of a rocky planet. Mars began to cool, volcanism ceased, atmosphere diminished. In other words, a limited inventory of radiogenic isotopes rather than the distance from the Sun has dramatically impacted the Mars evolution changing its geology and causing collapse of the atmosphere. Likewise Earth and Venus, Mars probably possessed an ocean, currently its significant fraction being preserved beneath the surface as underground water.

It is important to emphasize that synergy of the Earth and planetary sciences is aimed at a better understanding of the present, past, and future of the Earth based on comparative planetology approach. It is intended to answer the key questions about feedback for the existing regulation mechanisms to maintain clement climate. Concurrently, this contributes to the cardinal problems of planetary cosmogony allowing us to put more stringent constraints on the range of parameters used in modeling of the origin and evolution of the solar and other planetary systems.

Ring System

Planetary rings are not solid objects. They are rings of dust and other particles that are in orbit around a planet- almost like billions upon billions of tiny moons all orbiting close together so that from a distance the rings actually look like a hoop around a planet.

The most famous planetary rings are without a doubt those of Saturn, but the other gas giants in our solar system (Jupiter, Uranus and Neptune) also possess less impressive ring systems of their own. In 2008 the Cassini spacecraft found evidence of the possibility around Rhea, one of Saturn's moons. If this evidence is corroborated, Rhea will be the only body other than a planet to have a ring system.

What Causes Rings?

A ring is a collection of vast numbers of particles, each like a tiny moon obeying Kepler's laws as it follows its own orbit around the planet. Thus, the inner particles revolve faster than those farther out, and the ring as a whole does not rotate as a solid body. In fact, it is better not to think of a ring rotating at all, but rather to consider the revolution (or motion in orbit) of its individual moonlets.

If the ring particles were widely spaced, they would move independently, like separate moonlets. However, in the main rings of Saturn and Uranus the particles are close enough to exert mutual gravitational influence, and occasionally even to rub together or bounce off each other in low-speed collisions. Because of these interactions, we see phenomena such as waves that move across the rings just the way water waves move over the surface of the ocean.

There are two basic ideas of how such rings come to be. First is the breakup hypothesis, which suggests that the rings are the remains of a shattered moon. A passing comet or asteroid might have collided with the moon, breaking it into pieces. Tidal forces then pulled the fragments apart, and they dispersed into a disk. The second hypothesis, which takes the reverse perspective, suggests that the rings are made of particles that were unable to come together to form a moon in the first place.

In either theory, the gravity of the planet plays an important role. Close to the planet, tidal forces can tear bodies apart or inhibit loose particles from coming together. We do not know which explanation holds for any given ring, although many scientists have concluded that at least a few of the rings are relatively young and must therefore be the result of breakup.

Four Ring Systems

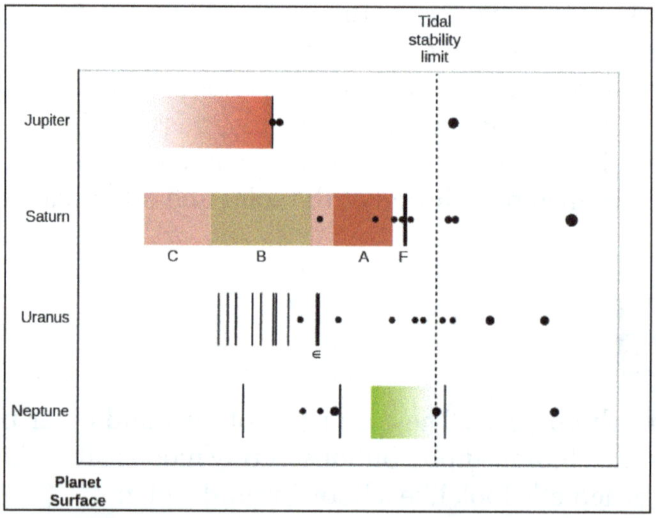

This diagram shows the locations of the ring systems of the four giant planets.

The left axis represents the planet's surface. The dotted vertical line is the limit inside which gravitational forces can break up moons (each planet's system is drawn to a different scale, so that this stability limit lines up for all four of them). The black dots are the inner moons of each planet on the same scale as its rings. Notice that only really small moons survive inside the stability limit.

Rings of Saturn

Saturn's rings are one of the most beautiful sights in the solar system. From outer to inner, the three brightest rings are labeled with the extremely unromantic names of A, B, and C Rings. Table gives the dimensions of the rings in both kilometers and units of the radius of Saturn, R_{Saturn}. The B Ring is the brightest and has the most closely packed particles, whereas the A and C Rings are translucent.

The total mass of the B Ring, which is probably close to the mass of the entire ring system, is about equal to that of an icy moon 250 kilometers in diameter (suggesting that the ring could have originated in the breakup of such a moon). Between the A and B Rings is a wide gap named the Cassini Division after Gian Domenico Cassini, who first glimpsed it through a telescope in 1675 and whose name planetary scientists have also given to the Cassini spacecraft exploring the Saturn system.

Saturn's Rings as Seen from Above and Below.

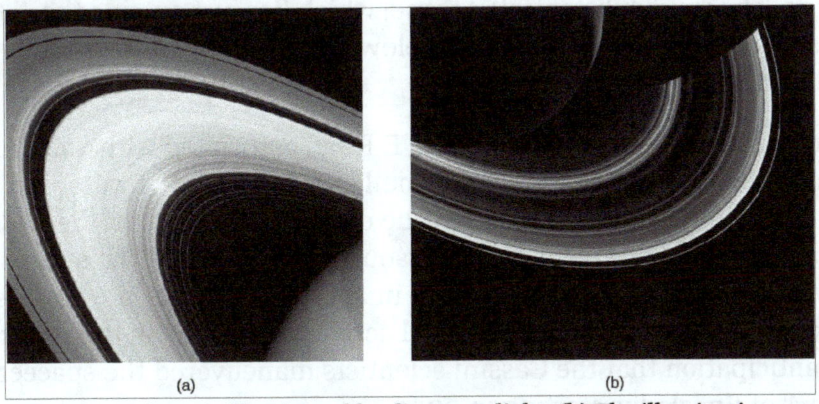

(a) The view from above is illuminated by direct sunlight. (b) The illumination seen from below is sunlight that has diffused through gaps in the rings.

Selected Features in the Rings of Saturn			
Ring Name[1]	Outer Edge (R_{Saturn})	Outer Edge (km)	Width (km)
F	2.324	140,180	90
A	2.267	136,780	14,600
Cassini Division	2.025	122,170	4590
B	1.949	117,580	25,580
C	1.525	92,000	17,490

Saturn's rings are very broad and very thin. The width of the main rings is 70,000 kilometers, yet their average thickness is only 20 meters. If we made a scale model of the rings out of paper, we would have to make them 1 kilometer across. On this scale, Saturn itself would loom as high as an 80-story building. The ring particles are composed primarily of water ice, and they range from grains the size of sand up to house-sized boulders. An insider's view of the rings would probably resemble a bright cloud of floating snowflakes and hailstones, with a few snowballs and larger objects, many of them loose aggregates of smaller particles.

Artist's Idealized Impression of the Rings of Saturn as Seen from the Inside.

Note that the rings are mostly made of pieces of water ice of different sizes. At the end of its mission, the Cassini spacecraft is planning to cut through one of the gaps in Saturn's rings, but it won't get this close.

In addition to the broad A, B, and C Rings, Saturn has a handful of very narrow rings no more than 100 kilometers wide. The most substantial of these, which lies just outside the A Ring, is called the F Ring; its surprising appearance is discussed below. In general, Saturn's narrow rings resemble the rings of Uranus and Neptune.

There is also a very faint, tenuous ring, called the E Ring, associated with Saturn's small icy moon Enceladus. The particles in the E Ring are very small and composed of water ice. Since such a tenuous cloud of ice crystals will tend to dissipate, the ongoing existence of the E Ring strongly suggests that it is being continually replenished by a source at Enceladus. This icy moon is very small only 500 kilometers in diameter but the Voyager images showed that the craters on about half of its surface have been erased, indicating geological activity sometime in the past few million years. It was with great anticipation that the Cassini scientists maneuvered the spacecraft orbit to allow multiple close flybys of Enceladus starting in 2005.

Those awaiting the Cassini flyby results were not disappointed. High-resolution images showed long, dark stripes of smooth ground near its south pole, which were soon nicknamed "tiger stripes". Infrared measurements revealed that these tiger stripes are warmer than their surroundings. Best of all, dozens of cryovolcanic vents on the tiger stripes were seen to be erupting geysers of salty water and ice. Estimates suggested that 200 kilograms of material were shooting into space each second not a lot, but enough for the spacecraft to sample.

Enceladus

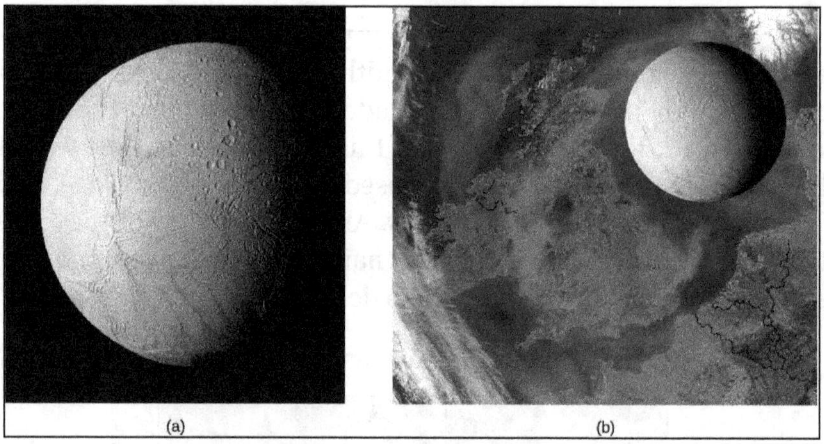

(a) This image shows both smooth and cratered terrain on Saturn's moon, and also "tiger stripes" in the south polar region (lower part of image). (b) Here Enceladus is shown to scale with Great Britain and the coast of Western Europe, to emphasize that it is a small moon, only about 500 kilometers in diameter.

These dark stripes (shown here in exaggerated color) have elevated temperatures and are the source of the many geysers discovered on Enceladus. They are about 130 kilometers long and 40 kilometers apart.

When Cassini was directed to fly into the plumes, it measured their composition and found them to be similar to material we see liberated from comets. The vapor and ice plumes consisted mostly of water, but with trace amounts of nitrogen, ammonia, methane, and other hydrocarbons. Minerals found in the geysers in trace amounts included ordinary salt, meaning that the geyser plumes were high-pressure sprays of salt water.

Based on the continuing study of Enceladus' bulk properties and the ongoing geysers, in 2015 the Cassini mission scientists tentatively identified a subsurface ocean of water feeding the geysers. These discoveries suggested that in spite of its small size, Enceladus should be added to the list of worlds that we would like to explore for possible life. Since its subsurface ocean is conveniently escaping into space, it might be much easier to sample than the ocean of Europa, which is deeply buried below its thick crust of ice.

Geysers on Enceladus

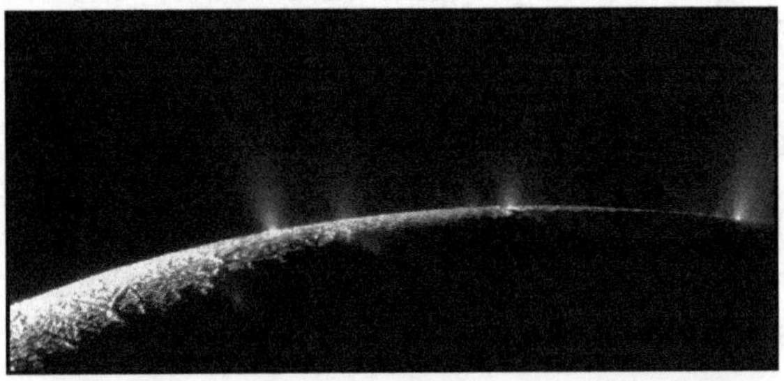

This Cassini image shows a number of water geysers on Saturn's small moon Enceladus, apparently salty water from a subsurface source escaping through cracks in the surface. You can see curved lines of geysers along the four "tiger stripes" on the surface.

Rings of Uranus and Neptune

Uranus' rings are narrow and black, making them almost invisible from Earth. The nine main rings were discovered in 1977 from observations made of a star as Uranus passed in front of it. We call such a passage of one astronomical object in front of another an occultation. During the 1977 occultation, astronomers expected the star's light to disappear as the planet moved across it. But in addition, the star dimmed briefly several times before Uranus reached it, as each narrow ring passed between the star and the telescope. Thus, the rings were mapped out in detail even though they could not be seen or photographed directly, like counting the number of cars in a train at night by watching the blinking of a light as the cars successively pass in front of it. When Voyager approached Uranus in 1986, it was able to study the rings at close range; the spacecraft also photographed two new rings.

Rings of Uranus

The Voyager team had to expose this image for a long time to get a glimpse of Uranus' narrow dark rings. You can see the grainy structure of "noise" in the electronics of the camera in the picture background.

The outermost and most massive of Uranus' rings is called the Epsilon Ring. It is only about 100 kilometers wide and probably no more than 100 meters thick (similar to the F Ring of Saturn). The Epsilon Ring encircles Uranus at a distance of 51,000 kilometers, about twice the radius of Uranus. This ring probably contains as much mass as all of Uranus' other ten rings combined; most of them are narrow ribbons less than 10 kilometers wide, just the reverse of the broad rings of Saturn.

The individual particles in the uranian rings are nearly as black as lumps of coal. While astronomers do not understand the composition of this material in detail, it seems to consist in large part of carbon and hydrocarbon compounds. Organic material of this sort is rather common in the outer solar system. Many of the asteroids and comets are also composed of dark, tarlike materials. In the case of Uranus, its ten small inner moons have a similar composition, suggesting that one or more moons might have broken up to make the rings.

Neptune's rings are generally similar to those of Uranus but even more tenuous. There are only four of them, and the particles are not uniformly distributed along their lengths. Because these rings are so difficult to investigate from Earth, it will probably be a long time before we understand them very well.

Rings of Neptune

This long exposure of Neptune's rings was photographed by Voyager 2. Note the two denser regions of the outer ring.

Interactions between Rings and Moons

Much of our fascination with planetary rings is a result of their intricate structures, most of which owe their existence to the gravitational effect of moons, without which the rings would be flat and featureless. Indeed, it is becoming clear that without moons there would probably be no rings at all because, left to themselves, thin disks of small particles gradually spread and dissipate.

Most of the gaps in Saturn's rings, and also the location of the outer edge of the A Ring, result from gravitational resonances with small inner moons. A resonance takes place when two objects have orbital periods that are exact ratios of each other, such as 1:2 or 2:3. For example, any particle in the gap at the inner side of the Cassini Division of Saturn's rings would have a period equal to one-half that of Saturn's moon Mimas. Such a particle would be nearest Mimas in the same part of its orbit every second revolution. The repeated gravitational tugs of Mimas, acting always in the same direction, would perturb it, forcing it into a new orbit outside the gap. In this way, the Cassini Division became depleted of ring material over long periods of time.

The Cassini mission revealed a great deal of fine structure in Saturn's rings. Unlike the earlier Voyager flybys, Cassini was able to observe the rings for more than a decade, revealing a remarkable range of changes, on time scales from a few minutes to several years. Many of the features newly seen in Cassini data indicated the presence of condensations or small moons only a few tens of meters across imbedded in the rings. As each small moon moves, it produces waves in the surrounding ring material like the wake left by a moving ship. Even when the moon is too small to be resolved, its characteristic waves could be photographed by Cassini.

One of the most interesting rings of Saturn is the narrow F Ring, which contains several apparent ringlets within its 90-kilometer width. In places, the F Ring breaks up into two or three parallel strands that sometimes show bends or kinks. Most of the rings of Uranus and Neptune are also narrow ribbons like the F Ring of Saturn. Clearly, the gravity of some objects must be keeping the particles in these thin rings from spreading out.

As we have seen, the largest features in the rings of Saturn are produced by gravitational resonances with the inner moons, while much of the fine structure is caused by smaller embedded moons. In the case of Saturn's F Ring, close-up images revealed that it is bounded by the orbits of two moons, called Pandora and Prometheus. These two small moons (each about 100 kilometers in diameter) are referred to as shepherd moons, since their gravitation serves to "shepherd" the ring particles and keep them confined to a narrow ribbon. A similar situation applies to the Epsilon Ring of Uranus, which is shepherded by the moons Cordelia and Ophelia. These two shepherds, each about 50 kilometers in diameter, orbit about 2000 kilometers inside and outside the ring.

Saturn's F Ring and its Shepherd Moons

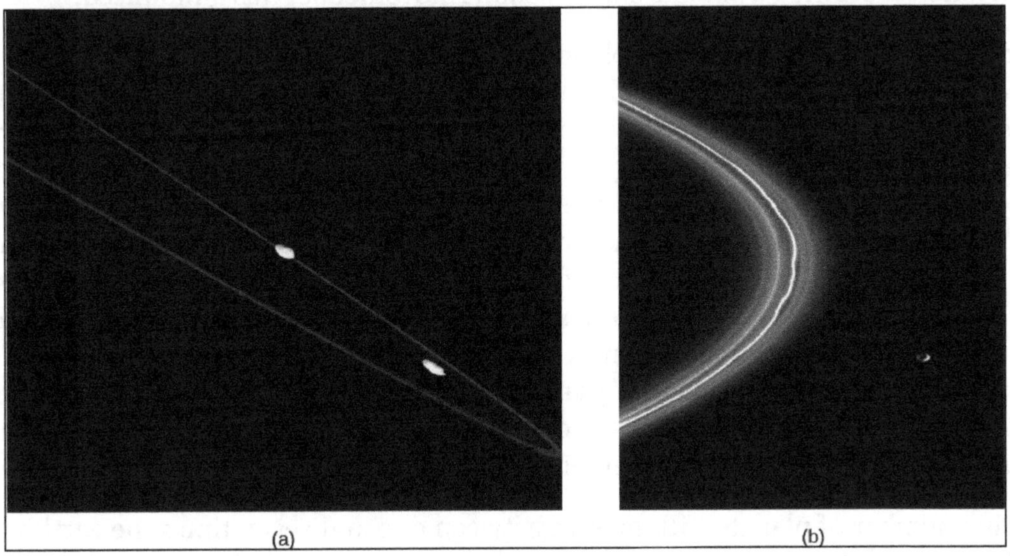

(a) This Cassini image shows the narrow, complex F Ring of Saturn, with its two small shepherd moons Pandora (left) and Prometheus (right). (b) In this closer view, the shepherd moon Pandora (84 kilometers across) is seen next to the F ring, in which the moon is perturbing the main (brightest) strand of ring particles as it passes. You can see the dark side of Pandora on this image because it is being illuminated by the light reflected from Saturn.

Theoretical calculations suggest that the other narrow rings in the uranian and neptunian systems should also be controlled by shepherd moons, but none has been located. The calculated diameter for such shepherds (about 10 kilometers) was just at the limit of detectability for the Voyager cameras, so it is impossible to say whether they are present or not.

One of the outstanding problems with understanding the rings is determining their ages. Have the giant planets always had the ring systems we see today, or might these be a recent or transient addition to the solar system? In the case of the main rings of Saturn, their mass is about the same as that of the inner moon Mimas. Thus, they could have been formed by the break-up of a Mimas-sized moon, perhaps very early in solar system history, when there were many interplanetary projectiles left over from planet formation. It is harder to understand how such a catastrophic event could have taken place recently, when the solar system had become a more stable place.

Planetary System

planetary system, a star and all the celestial bodies bound to it by gravity, especially planets and their natural satellites . Until the last decade of the 20th cent., the only planetary system known was the solar system , which comprises the sun and the surrounding planets, natural satellites, asteroids, meteoroids, comets, and other celestial bodies. Speculation that other planetary systems exist dates back to antiquity, and through the years ever increasing numbers of astronomers searched for earthlike planets circling sunlike stars. The breakthrough came in 1992, when radio astronomers detected three planets orbiting a pulsar however, because pulsars are not normal stars, this was not considered a true planetary system. The first detection of an extrasolar planet, or exoplanet, around a normal star, 51 Pegasi, was made in 1995. This was quickly followed by the detection of a number of single exoplanets orbiting normal stars, and in 1999 the first discovery of multiple exoplanets orbiting a sunlike star, Upsilon Andromedae, was announced.

Because the solar system was the only planetary system known, all models of planetary systems were based on its characteristics several small planets close to the star, several large planets at greater distances, and circular planetary orbits. Most of the extrasolar planets discovered so far, however, are larger than earth, and many of those are much larger than Jupiter, the largest of the solar planets many of the rocky planets are much larger than earth, often up to 10 times more massive (one Kepler 10c, is 17 times as massive) many orbit their star at distances less than that of Mercury, the solar planet closest to the sun in one system found by the *Kepler* space telescope, five planets orbit a star more closely than Mercury does the sun and many have elliptical rather than circular orbits. Planets have also been found orbiting binary stars. All of this has caused planetary scientists to revisit accepted theories of planetary formation. Future theories will be measured against stars surrounded by a ring of gas and dust, such as Beta Pictoris, which are thought to be young adult stars with a planetary system forming around them.

An increasing number of planets with masses between one and seven times the earth's have been found in the 21st cent. In 2013 it was announced that astronomers reviewing the data from the *Kepler* space telescope had identified a planetary system (Kepler 37) that included a planet (Kepler 37c) somewhat smaller than the earth and another (Kepler 37b) that was smaller than Mercury. Kepler 37b is the smallest extrasolar planet discovered so far neither of the two planets was in the habitable zone. In 2014, Kepler scientists announced the discovery of a habitable-zone planet (Kepler 186f) with a radius estimated to be 10% larger than the earth's, that orbited a cool dwarf star with four other planets because of its size, Kepler 186f was believed to be a rocky planet with the potential to have liquid water. It is possible that some of the bodies that have been discovered are not planets in the solar-system sense but a new class of celestial bodies or even brown dwarfs .

Because stars are so distant and bright and an extrasolar planet, no matter how large, is relatively small and dim, it cannot be seen or photographed directly in visible light. Several techniques have been used to infer the presence of such planets. *Astrometry* is based on the slight gravitational disturbance, or wobble, that the planet causes in the motion of the star. *Photometry,* also called the *transit method,* is to measure the distinct dimming of light from the star as the planet's orbit brings it between the star and the earth. Using photometric techniques it also has been possible to photograph extrasolar planets in infrared light. *Doppler spectroscopy* is based on the fact that a planet periodically pulls its star closer to and farther from the earth as it orbits the star this motion has a measurable effect on the spectrum of light coming from the star. In *pulsar timing,* planets orbiting a pulsar can be detected by measuring the periodic variation in the pulse arrival time however, because the planets are orbiting a pulsar, a dead star, rather than a main-sequence star like the sun, this tends to be of less interest in the search for an earthlike extrasolar planet.

Heliocentrism

Historically, heliocentrism (that the sun was the centre of the universe) was opposed to geocentrism, which placed the Earth at the centre of the universe. The notion of a heliocentric solar system, with the Sun at the centre, is possibly first suggested in the Vedic literature of ancient India, which often refer to the Sun as the "centre of spheres". Some interpret Aryabhatta's writings in Āryabhaṭīya as implicitly heliocentric. The idea was first proposed in western philosophy and Greek astronomy as early as the 3rd century BC by Aristarchus of Samos, but had received no support from most other ancient astronomers.

Speculation on Extrasolar Planetary Systems

In the 16th century the Italian philosopher Giordano Bruno, an early supporter of the Copernican theory that the Earth and other planets orbit the Sun, put forward the view that the fixed stars are similar to the Sun and are likewise accompanied by planets. He was burned at the stake for his ideas by the Roman Inquisition.

In the 18th century the same possibility was mentioned by Isaac Newton in the "General Scholium" that concludes his Principia. Making a comparison to the Sun's planets, he wrote "And if the fixed stars are the centres of similar systems, they will all be constructed according to a similar design and subject to the dominion of One. His theories gained traction through the 19th and 20th centuries despite a lack of supporting evidence. Long before their confirmation by astronomers, conjecture on the nature of planetary systems had been a focus of Search for extraterrestrial intelligence and has been a prevalent theme in fiction, particularly science fiction.

Detection of Exoplanets

The first confirmed detection of a planetary system was in 1992, with the discovery of several terrestrial-mass planets orbiting the pulsar PSR B1257+12. The first confirmed detection of planetary system of a main-sequence star was made in 1995, when a giant planet, 51 Pegasi b, was found in a four-day orbit around the nearby G-type star 51 Pegasi. The frequency of detections has increased since then, particularly through advancements in methods of detecting extrasolar planets and dedicated planet finding programs such as the Kepler mission.

An artist's concept of a protoplanetary disk. Image from Smithsonian.

Origin and Evolution

Early theories of planetary systems were based on knowledge of the formation and evolution of the Solar System. Planetary systems are generally believed to form as part of the same process which results in star formation. Some early theories involved another star passing extremely close to the star, drawing material out from it which then coalesced to form the planets. However, the probability of such a near collision is now known to be far too low to make this a viable model. Accepted theories today argue that a protoplanetary disk forms by gravitational collapse of a molecular cloud and then evolves into a planetary system by collisions and gravitational capture.

Some planetary systems may form differently, however. Planets orbiting pulsars stars which emit periodic bursts of electromagnetic radiation have been discovered by the slight variations they cause in the timing of these bursts. Pulsars are formed in violent supernova explosions, and a normal planetary system could not possibly survive such a blast planets would either evaporate, be pushed off of their orbits by the masses of gas from the exploding star, or the sudden loss of most of the mass of the central star would see them escape the gravitational hold of the star. One theory is that existing stellar companions were almost entirely evaporated by the supernova blast, leaving behind planet-sized bodies. Alternatively, planets may somehow form in the accretion disk surrounding pulsars.

The formation of planetary systems has also been found to be linked to stellar classification. Stars are composed mainly of the light elements hydrogen and helium. They also contain a small proportion of heavier elements such as iron, and this fraction is referred to as a star's metallicity. Stars of higher metallicity are much more likely to form and retain planetary systems, and the planets within them tend to be more massive than those of lower-metallicity stars. It has also been shown that stars with planets are more likely to be deficient in lithium. Recent observations by the Spitzer Space Telescope indicate that planetary formation does not occur around other stars in the vicinity of an O class star due to the photoevaporation effect.

Models of the Solar System point to material being ejected during the formation and early evolution as the system stabilised. Similar interactions are thought to be common to many planetary systems. The orbits of large bodies are being established; they may migrate and cause planetary body collisions. Others may be sufficiently perturbed to be ejected completely from the system. These ejected bodies are known as rogue planets and may retain their natural satellites. Systems such as these in interstellar space, known as planetars, may orbit the galaxy directly and although difficult to detect are now thought to be extremely common.

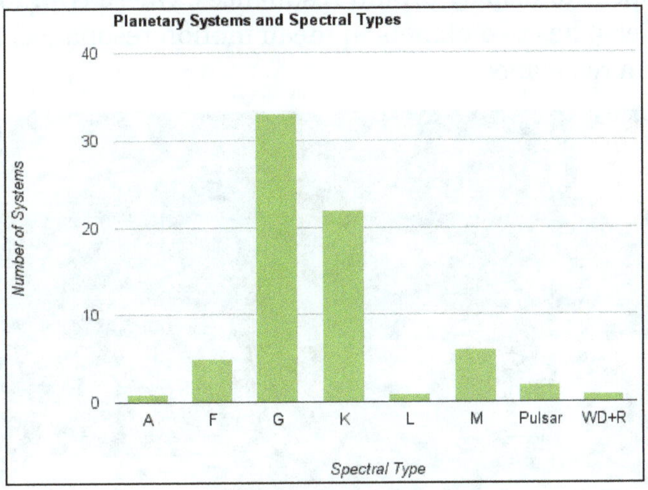

Planetary systems.

Types and Attributes

With the increasing discovery of planetary systems, astronomers have begun to classify them by type. In particular, as certain types of stars are now known for producing specific types of planetary systems, they are classified by the spectral type of the host star. Main-sequence stars like the Sun (spectral type G or K), for example, have represented the majority of planetary system discoveries. Additionally, they are often classified by the size and types of planets and their orbital configuration. Hot Jupiter systems with a gas giant very close to the star are one of the most common found so far and Hot Neptune type systems have also been found. Theories, such as scattering, have been proposed for the formation of large planets close to their parent stars. Dusty disks with a wide ring of dust and comets are another common type of system. Protoplanetary disks, still in the process of forming have also been discovered. At present, few systems have been found to analogues of our own with terrestrial planets close to the parent star. More commonly, systems consisting of multiple Super-Earths have been detected.

Multiplanet Systems

140 systems (including the Solar System) are now known to consist of two or more planets. In terms of total confirmed planets, the Solar System with 8 remains the largest planetary system and HD 10180 with 7, the second largest. 55 Cancri was for some time the largest known system. Some systems, such as HD 10180 and Gliese 581 contain multiple unconfirmed planets pushing the total expected number of planets to be discovered in their systems higher. If HD 10180's two candidate planets are confirmed, it would be the largest by number of known planets with a total of 9.

Orbital Properties

Unlike the Solar System, which has coplanar and circular orbits, many of the known planetary systems display much higher orbital eccentricity and some consist of planets in dramatic elliptical orbits while some consist of planets on inconsistent orbital inclination. An example of such a system is 16 Cygni. Although no planets in the Solar System are completely tidally locked, planets with regular orbits in close to their primary stars are thought to typically experience tidal locking. aaSome systems also appear to display orbital resonances. Gliese 876, for example has a 3-body Laplace resonance. Kepler-9 has two planets in mean motion resonance, the first transiting planets discovered with such a resonance.

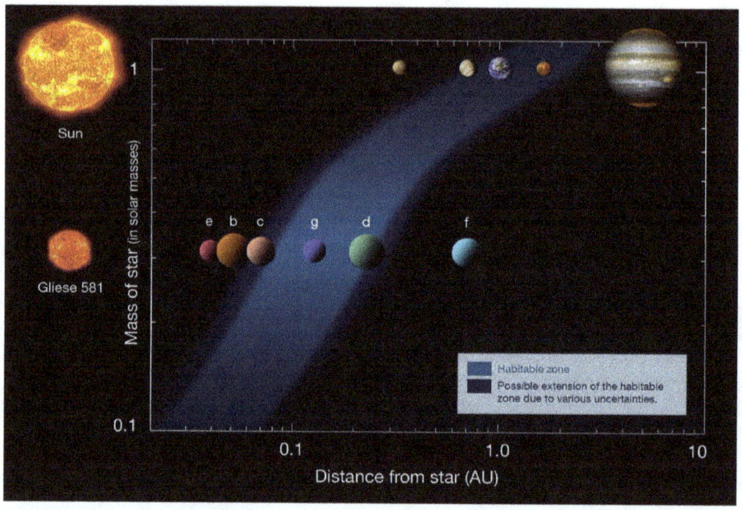

The habitable zone of Gliese 581 compared with our Solar System's habitable zone.

Habitable Zones

The habitable zone is the region in the planetary system where an Earth-like planet (or moon) could maintain liquid water on its surface and potentially therefore, life. The distance of this zone from the parent star can vary greatly from system to system. The lack of surface water on Mars, Venus, the Moon, and Ceres, which according to some estimates lie within the habitable zone of the Solar System, does not exclude the possibility of extraterrestrial surface life in the Solar System but makes it highly unlikely. However the abundance of Life on Earth makes the habitable zones of exoplanetary systems of particular interest to exobiologists and the Search for extraterrestrial intelligence. Gliese 581 was the first exoplanetary system found with Super-Earths in the habitable zone. HD 37124 has multiple known planets in the zone. Systems with gas giants have more commonly discovered in the habitable zone, though they could possibly harbour Earth-like natural satellites.

Main Sequence Systems

The Solar System was the first planetary system known around a main-sequence star. As such, main-sequence systems are of particular interest to science due to the prospect of finding true analogs to the Solar System, along with true Earth analogs and therefore possibly Earth-like complex or even intelligent life.

The first main-sequence exoplanetary system found was Upsilon Andromedae in April 1999. Since then, several further systems have been discovered. Many have been found to bear similarities to our own, for example the first terrestrial planets of comparable size to Earth (in the Kepler-20 system) and planets close to Earth sized but much larger (Super-Earths) have even been located in habitable zones in the HD 85512 and Kepler 22 systems. Discoveries to date indicate that our Solar System's arrangement of all the terrestrial planets arranged closest to the sun and the giant planets further out is unlikely to be the general rule. This challenges well established theories of their formation and evolution suggesting that early migration of giant planets could causes wide variations in the arrangement of main-sequence systems.

Three of the nearest main-sequence stars to the Sun are now thought to host planetary systems. This includes the nearest main-sequence star to Earth at 4.37ly, Alpha Centauri A which was found in 2012 to have an Earth-sized planet in close orbit around it. Unconfirmed planetary systems include the Epsilon Eridani system at 10.5ly from Earth and the Tau Ceti system at 11.905ly, both believed to have multiple detected planets.

This artist's concept illustrates a young, red dwarf star surrounded by three planets.

Red Dwarf Systems

Red dwarf systems orbit red dwarfs, stars which may represent 70% to more than 90% of the stars in the galaxy. The Gliese 876 was the first such system discovered in 1998. Most planets of these systems are thought to be tidally locked. This and high stellar variation are postulated to be major impediments to planetary habitability. Gliese 581 is another well known red dwarf system which is speculated to have multiple planets within the habitable zone.

Brown Dwarf Systems

Brown dwarf planetary system orbit brown dwarfs. 2M1207, the first planetary system to be imaged, was also the first such system discovered. Some companions of brown dwarfs, such as those of 2M1207 and GQ Lupi, are thought to have formed by cloud collapse rather than accretion, and hence these are sometimes considered sub-brown dwarfs rather than planets. Disks around brown dwarfs have been found to have many of the same features as disks around stars; therefore, it is expected that there will be accretion-formed planets around brown dwarfs. Given the small mass of brown dwarf disks, most planets will be terrestrial planets rather than gas giants. It is now believed that many brown dwarfs may have planets and also, possibly, a shortly lived habitable zone. Most planets of these systems are thought to be tidally locked.

Sub-brown Dwarf Systems

Rogue systems are systems that are not gravitationally bound to any star and that therefore orbit the galaxy directly. Some astronomers have estimated that there may be twice as many Jupiter-sized rogue planets as there are stars. Isolated planetary-mass objects which were not ejected from a star's orbit, may have formed in a similar way to stars. The IAU has proposed that those objects be called sub-brown dwarfs. It is theorised that objects ejected from a planetary system could retain a system of satellites. Cha 110913-773444 is an example of a rogue system with a protoplanetary disk speculated to contain as many as 4 satellites. It is unknown whether it is ejected planet, or formed on its own.

Pulsar Systems

Pulsar systems orbit pulsars, or rapidly rotating neutron stars. PSR B1257+12 was not only the first pulsar exoplanetary system but also the first planetary system discovered. PSR B1620-26 is also confirmed, however many other pulsar systems remain unconfirmed. The formation of pulsar systems is still the subject of research. It is thought that planets form from the debris of companion stars.

An artist's impression of the binary star system HD 98800 B, which is surrounded by a disc that may be in the process of forming planets. HD 98800 B is itself a member of a quadruple star system.

Multiple Star Systems

Multiple star systems, including binary stars are abundant in the Milky Way Galaxy and several are now known to possess planetary systems. The discovery of such systems however creates new challenges on theories on planetary system formation. Prior to the groundbreaking 1993 discovery of the first known planet PSR B1620-26 b in the binary system PSR B1620-26, the possibility of the existence of such systems was long theorised and debated. Chinese-American astrophysicist Su-Shu Huang had dampened expectations of planets forming or remaining in such systems in his 1959 paper "Life-Supporting Regions in the Vicinity of Binary Systems" claiming that due to orbital instability planets and in particular habitable planets would likely be rare. The film Star Wars, however captured the imagination of popular culture with its "twin sun" scene on Tatooine.

PSR B1620-26 was not only the first binary system (and multistar system) found with a planet but the first found in which the planet orbited both stars (i.e. in circumbinary orbit, a circumbinary planet).

16 Cygni, discovered in 1996 was the first confirmed triple star planetary system found however the planet orbits just one of the stars – 16 Cygni B. In 2011, the Kepler-16 system was found to have a circumbinary planet within its habitable zone, leading to the Smithsonian Centre informally referring to it as "Tatooine" in reference to Star Wars.

This type of planetary system is of particular interest due to Alpha Centauri A being the closest main-sequence star to Earth but also a binary (Alpha Centauri AB) contained within a triple star system. The notion of planets in the Centauri system has been a popular theme in fiction. In 2012 Alpha Centauri B was found to have an Earth-sized planet in orbit around it, increasing expectations of more terrestrial planet discoveries in this and similar systems. PH1, confirmed in 2012, is the first quadruple star planetary system confirmed. NN Serpentis and HW Virginis are notable multi-planet multi-star systems.

Transiting Systems

Transiting planetary systems are those with multiple planets in orbital planes along the line of sight between their star and Earth. The transit method is one of the most successful Earth-centric observation methods. The Kepler mission was designed to discover systems of this type and has found numerous candidate systems. Kepler-11 is a prominent example of a transiting system in which multiple simultaneous planetary transits have been observed. Transiting systems can be studied to not only discover planets, but through Astronomical spectroscopy and Transit timing effects accurately analyse the properties of the systems including the detection of extraterrestrial atmospheres, dust or asteroid belts and possibly natural satellites.

System Architectures

The Solar System consists of an inner region of small rocky planets and outer region of large gas giants. However, other planetary systems can have quite different architectures. Studies suggest that architectures of planetary systems are dependent on the conditions of their initial formation. Many systems with a hot Jupiter gas giant very close to the star have been found. Theories, such as planetary migration or scattering, have been proposed for the formation of large planets close to their parent stars. At present, few systems have been found to be analogous to the Solar System with terrestrial planets close to the parent star. More commonly, systems consisting of multiple Super-Earths have been detected.

Components

Planets

Some studies suggest that there is at least one planet on average per star. This would suggest that, like the Solar System, most stars have planets (or exoplanets). However, the proportion of stars is uncertain because not all planets can yet be detected. The radial-velocity method and the transit method (which between them are responsible for the vast majority of detections) are most sensitive to large planets in small orbits. Thus, many known exoplanets are "hot Jupiters": planets of Jovian mass or larger in very small orbits with periods of only a few days. A 2005 survey of radial-velocity-detected planets found that about 1.2% of Sun-like stars have a hot Jupiter, where "Sun-like star" refers to any main-sequence star of spectral classes late-F, G, or early-K without a close stellar companion. This 1.2% is more than double the frequency of hot jupiters detected by the Kepler spacecraft, which may be because the Kepler field of view covers a different region of the Milky Way where the metallicity of stars is different. It is further estimated that 3% to 4.5% of Sun-like stars possess a giant planet with an orbital period of 100 days or less, where "giant planet" means a planet of at least 30 Earth masses.

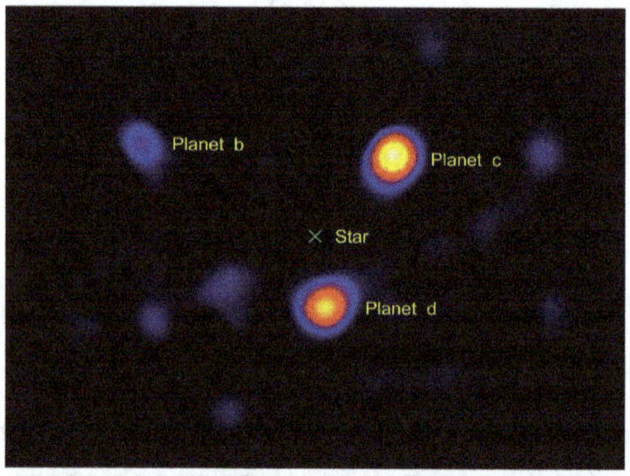

The three known planets of the star HR8799, as imaged by the Hale Telescope. The light from the central star was blanked out by a vector vortex coronagraph.

It is known that small planets (of roughly Earth-like mass or somewhat larger) are more common than giant planets. It also appears that there are more planets in large orbits than in small orbits. Based on this, it is estimated that perhaps 20% of Sun-like stars have at least one giant planet, whereas at least 40% may have planets of lower mass. A 2012 study of gravitational microlensing data collected between 2002 and 2007 concludes the proportion of stars with planets is much higher and estimates an average of 1.6 planets orbiting between 0.5–10 AU per star in the Milky Way. The authors of this study conclude that "stars are orbited by planets as a rule, rather than the exception".

Whatever the proportion of stars with planets, the total number of exoplanets must be very large. Because the Milky Way has at least 200 billion stars, it must also contain tens or hundreds of billions of planets.

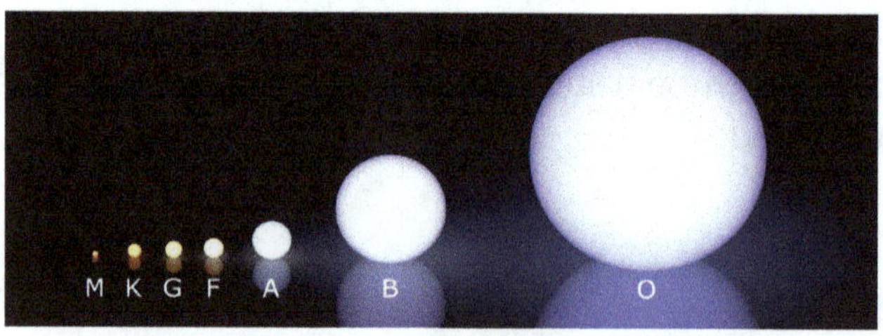

The Morgan-Keenan spectral classification.

Most known exoplanets orbit stars roughly similar to the Sun, that is, main-sequence stars of spectral categories F, G, or K. One reason is that planet-search programs have tended to concentrate on such stars. In addition, statistical analyses indicate that lower-mass stars (red dwarfs, of spectral category M) are less likely to have planets massive enough to be detected by the radial-velocity method. Nevertheless, several tens of planets around red dwarfs have been discovered by the Kepler spacecraft by the transit method, which can detect smaller planets.

Stars of spectral categories A and B typically rotate very quickly, which makes it very difficult to measure the small Doppler shifts induced by orbiting planets because the spectral lines are very broad. However, this type of massive star eventually evolves into a cooler red giant that rotates more slowly and thus can be measured using the radial-velocity method. A few tens of planets have been found around red giants.

Observations using the Spitzer Space Telescope indicate that extremely massive stars of spectral category O, which are much hotter than the Sun, produce a photo-evaporation effect that inhibits planetary formation. When the O-type star goes supernova any planets that had formed would become free-floating due to the loss of stellar mass unless the natal kick of the resulting remnant pushes it in the same direction as an escaping planet. Fallback disks of matter that failed to escape orbit during a supernova may form planets around neutron stars and black holes.

Doppler surveys around a wide variety of stars indicate about 1 in 6 stars having twice the mass of the Sun are orbited by one or more Jupiter-sized planets, vs. 1 in 16 for Sun-like stars and only 1 in 50 for red dwarfs. On the other hand, microlensing surveys indicate that long-period Neptune-mass planets are found around 1 in 3 red dwarfs. Kepler Space Telescope observations of planets with up to one year periods show that occurrence rates of Earth- to Neptune-sized planets (1 to 4 Earth radii) around M, K, G, and F stars are successively higher towards cooler, less massive stars.

At the low-mass end of star-formation are sub-stellar objects that don't fuse hydrogen: the brown dwarfs and sub-brown dwarfs, of spectral classification L,T and Y. Planets and protoplanetary disks have been discovered around brown dwarfs, and disks have been found around sub-brown dwarfs (e.g. OTS 44).

Ordinary stars are composed mainly of the light elements hydrogen and helium. They also contain a small proportion of heavier elements, and this fraction is referred to as a star's metallicity (even if the elements are not metals in the traditional sense), denoted [m/H] and expressed on a logarithmic scale where zero is the Sun's metallicity.

A 2012 study of the Kepler spacecraft data found that smaller planets, with radii smaller than Neptune's were found around stars with metallicities in the range $-0.6 < [m/H] < +0.5$ (about four times less than that of the Sun to three times more), whereas larger planets were found mostly around stars with metallicities at the higher end of this range (at solar metallicity and above). In this study small planets occurred about three times as frequently as large planets around stars of metallicity greater than that of the Sun, but they occurred around six times as frequently for stars of metallicity less than that of the Sun. The lack of gas giants around low-metallicity stars could be because the metallicity of protoplanetary disks affects how quickly planetary cores can form and whether they accrete a gaseous envelope before the gas dissipates. However, Kepler can only observe planets very close to their star and the detected gas giants probably migrated from further out, so a decreased efficiency of migration in low-metallicity disks could also partly explain these findings.

A 2014 study found that not only giant planets, but planets of all sizes, have an increased occurrence rate around metal-rich stars compared to metal-poor stars, although the larger the planet, the greater this increase as the metallicity increases. The study divided planets into three groups based on radius: gas giants, gas dwarfs, and terrestrial planets with the dividing lines at 1.7 and 3.9 Earth radii. For these three groups, the planet occurrence rates are 9.30, 2.03, and 1.72 times higher for metal-rich stars than for metal-poor stars, respectively. There is a bias against detecting smaller planets because metal-rich stars tend to be larger, making it more difficult to detect smaller planets, which means that these increases in occurrence rates are lower limits.

It has also been shown that stars with planets are more likely to be deficient in lithium.

Most stars form in open clusters, but very few planets have been found in open clusters and this led to the hypothesis that the open-cluster environment hinders planet formation. However, a 2011 study concluded that there have been an insufficient number of surveys of clusters to make such a hypothesis. The lack of surveys was because there are relatively few suitable open clusters in the Milky Way. Recent discoveries of both giant planets and low-mass planets in open clusters are consistent with there being similar planet occurrence rates in open clusters as around field stars. The open cluster NGC 6811 contains two known planetary systems Kepler-66 and Kepler-67.

Circumstellar Disks and Dust Structures

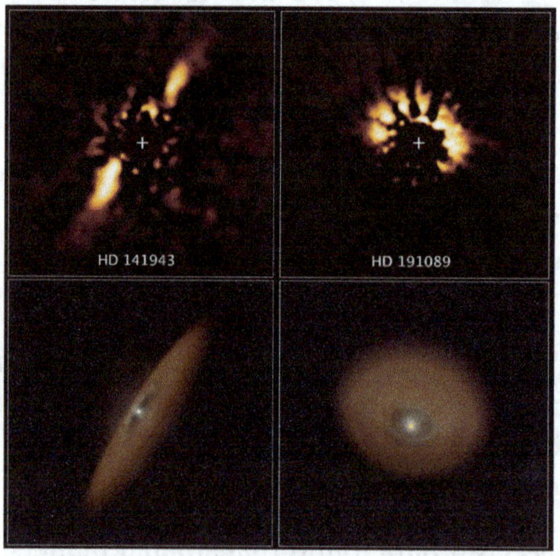

Debris disks detected in HST archival images of young stars, *HD 141943* and *HD 191089*, using improved imaging processes.

After planets, circumstellar disks are one of the most commonly observed properties of planetary systems, particularly of young stars. The Solar System possesses at least four major circumstellar disks (the asteroid belt, Kuiper belt, scattered disc, and Oort cloud) and clearly observable disks have been detected around nearby solar analogs including Epsilon Eridani and Tau Ceti. Based on observations of numerous similar disks, they are assumed to be quite common attributes of stars on the main sequence.

Interplanetary dust clouds have been studied in the Solar System and analogs are believed to be present in other planetary systems. Exozodiacal dust, an exoplanetary analog of zodiacal dust, the 1–100 micrometre-sized grains of amorphous carbon and silicate dust that fill the plane of the Solar System has been detected around the 51 Ophiuchi, Fomalhaut, Tau Ceti, and Vega systems.

Comets

There are 5,253 known Solar System comets and they are thought to be common components of planetary systems. The first exocomets were detected in 1987 around Beta Pictoris, a very young A-type main-sequence star. There are now a total of 11 stars around which the presence of exocomets have been observed or suspected. All discovered exocometary systems, are around very young A-type stars.

Other Components

Computer modeling of an impact in 2013 detected around the star NGC 2547 by the Spitzer Space Telescope and confirmed by ground observations suggests the involvement of large asteroids or protoplanets similar to the events believed to have led to the formation of terrestrial planets like the Earth.

Based on observations of the Solar System's large collection of natural satellites, they are believed common components of planetary systems; however, exomoons have so far eluded confirmation. The star 1SWASP J140747.93-394542.6, in the constellation Centaurus, is a strong candidate for a natural satellite. Indications suggest that the confirmed extrasolar planet WASP-12b also has at least one satellite.

Orbital Configurations

Unlike the Solar System, which has orbits that are nearly circular, many of the known planetary systems display much higher orbital eccentricity. An example of such a system is 16 Cygni.

Mutual Inclination

The mutual inclination between two planets is the angle between their orbital planes. Many compact systems with multiple close-in planets interior to the equivalent orbit of Venus are expected to have very low mutual inclinations, so the system (at least the close-in part) would be even flatter than the solar system. Captured planets could be captured into any arbitrary angle to the rest of the system. The only system where mutual inclinations have actually been measured is the Upsilon Andromedae system: the planets, c and d, have a mutual inclination of about 30 degrees.

Orbital Dynamics

Planetary systems can be categorized according to their orbital dynamics as resonant, non-resonant-interacting, hierarchical, or some combination of these. In resonant systems the orbital periods of the planets are in integer ratios. The Kepler-223 system contains four planets in an 8:6:4:3 orbital resonance. Giant planets are found in mean-motion resonances more often than smaller planets. In interacting systems the planets orbits are close enough together that they perturb the orbital parameters. The Solar System could be described as weakly interacting. In strongly interacting systems Kepler's laws do not hold. In hierarchical systems the planets are arranged so that the system can be gravitationally considered as a nested system of two-bodies, e.g. in a star with a close-in hot jupiter with another gas giant much further out, the star and hot jupiter form a pair that appears as a single object to another planet that is far enough out.

Other, as yet unobserved, orbital possibilities include: double planets; various co-orbital planets such as quasi-satellites, trojans and exchange orbits; and interlocking orbits maintained by precessing orbital planes.

Number of Planets, Relative Parameters and Spacings

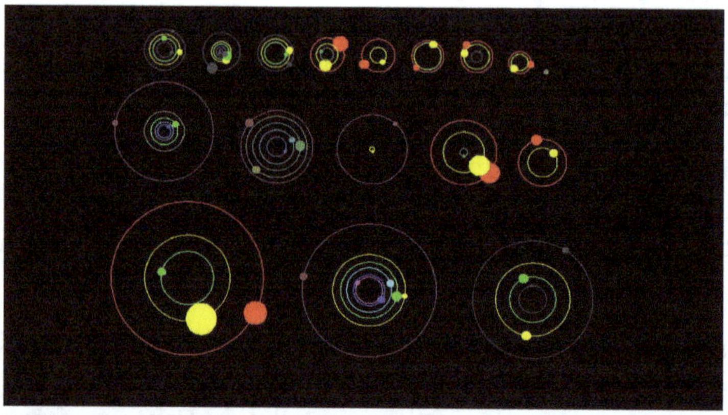

The spacings between orbits vary widely amongst the different systems discovered by the Kepler spacecraft.

- On The Relative Sizes of Planets Within Kepler Multiple Candidate Systems, David R. Ciardi.
- The Kepler Dichotomy among the M Dwarfs: Half of Systems Contain Five or More Coplanar Planets, Sarah Ballard, John Asher Johnson.
- Exoplanet Predictions Based on the Generalised Titius-Bode Relation, Timothy Bovaird, Charles H. Lineweaver.
- The Solar System and the Exoplanet Orbital Eccentricity - Multiplicity Relation, Mary Anne Limbach, Edwin L. Turner.
- The period ratio distribution of Kepler's candidate multiplanet systems, Jason H. Steffen, Jason A. Hwang.
- Are Planetary Systems Filled to Capacity? A Study Based on Kepler Results, Julia Fang, Jean-Luc Margot.

Planet Capture

Free-floating planets in open clusters have similar velocities to the stars and so can be recaptured. They are typically captured into wide orbits between 100 and 10^5 AU. The capture efficiency decreases with increasing cluster size, and for a given cluster size it increases with the host/primary mass. It is almost independent of the planetary mass. Single and multiple planets could be captured into arbitrary unaligned orbits, non-coplanar with each other or with the stellar host spin, or pre-existing planetary system. Some planet–host metallicity correlation may still exist due to the common origin of the stars from the same cluster. Planets would be unlikely to be captured around neutron stars because these are likely to be ejected from the cluster by a pulsar kick when they form. Planets could even be captured around other planets to form free-floating planet binaries. After the cluster has dispersed some of the captured planets with orbits larger than 10^6 AU would be slowly disrupted by the galactic tide and likely become free-floating again through encounters with other field stars or giant molecular clouds.

Small Solar System Body

Small body, also called small solar system body (SSSB), any natural solar system object other than the Sun and the major planets and dwarf planets and their satellites (moons). The small bodies populate the solar system in vast numbers and include the mostly rocky asteroids, or minor planets, the predominantly icy comets, and the fragments of such bodies commonly called meteoroids over a continuum of sizes down to microscopic grains known as interplanetary dust particles or micrometeoroids. The term small body is sometimes restricted to objects that can be observed telescopically while still outside Earth's atmosphere. In practice, that results in a lower limit of a few metres on the diameters of small bodies.

Composite image of the nucleus of Comet Halley.

The use of meteoroid has expanded to describe any small chunk of matter in interplanetary space, especially one less than a few tens of metres in size. The term formerly was limited to small bodies that were near enough to Earth to eventually enter its atmosphere. In that narrower meaning, meteoroid is usually linked with the terms meteor, meteorite, or both. As a meteoroid travels through the atmosphere, friction heats it up, creating a glowing trail of hot ionized gases (plasma) called a meteor. If the object survives both its passage through the atmosphere and its subsequent impact on the ground, it is called a meteorite. Over 99 percent of meteorites are fragments of asteroids. A small group is known to be of lunar origin, and a second group is generally acknowledged to have come from Mars. There is also reason to believe that some are fragments of the rocky remnants of comets, although that remains to be firmly established.

The Oort Cloud is a spherical shell of millions of icy bodies which surrounds the solar system at vast distances and is thought to be the birth place of long-period comets.

The Kuiper Belt is a region extending from Neptune's orbit out to the far and distant reaches of the solar system and possibly holds the best available record of the original interstellar materials that formed the solar nebula. This region beyond Neptune is also the most probable birthplace of the short-period comets.

Comets are pristine remnants from the formation of the solar system that are comprised of minerals, rock and mostly ice, much like a dirty snowball. They travel around the sun in elliptical orbits and can be inclined to the plane of the solar system at any angle. Comets can sprout tails extending many tens of millions of miles, during their closest approach to the sun. Short period comets are thought to come from the Kuiper Belt on the outskirts of Neptune's orbit and further, and longer period comets are thought to come from the Oort cloud, a vast spherical shell that surrounds the solar system at a huge distance. Recent spacecraft encounters with comets seem to raise more questions then they answer and some finds are quite unexpected. NASA targets some of these bodies with spacecraft loaded with instrumentation that help tease out the secrets lurking in these icy bodies.

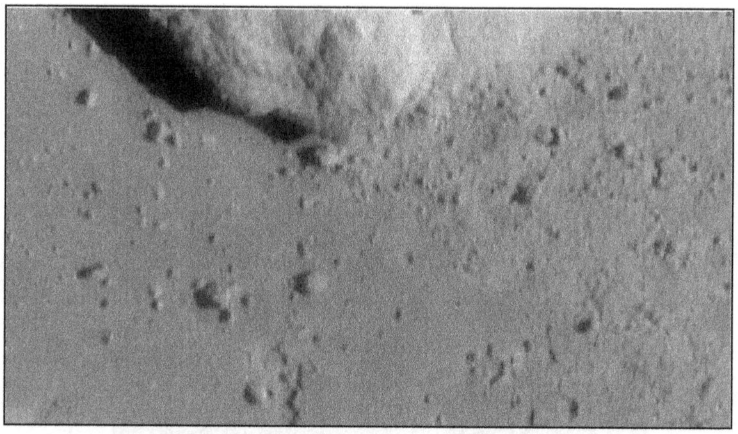

Pluto resides in the Kuiper Belt. With an orbit inclined to the plane of the solar system, Pluto most likely evolved away from the sun's flattened disk where the larger bodies (or planets) formed. Pluto's orbit crosses inside that of Neptune's rendering Pluto also a member of the Trans-Neptunian Object (TNO) class. The Pluto system is very exotic, having three moons including Charon discovered in 1978, and Nix and Hydra discovered in 2005.

Asteroids are rocky remnants from the formation of the solar system. They are not spherical and have differing compositions and histories. Most, although not all asteroids, reside in a region between Mars and Jupiter where numerous other small rocky worlds orbit the sun. Some asteroids belong to groups that came from larger parent bodies which were shattered in past collisions with other asteroids. Some are in orbits that cross paths with that of Earth's or other planets. Asteroids that cross Earth's orbit are called Potentially Hazardous Asteroids (PHA) and the more we observe the heavens, the more of them we find, some of which are seen for the first time just after passing close to Earth.

Populations

The largest known small bodies, in the conventional sense, are several icy Kuiper belt objects found orbiting the Sun beyond the orbit of Neptune. Ceres which is the largest main-belt asteroid and is now considered a dwarf planet is roughly 950 km (590 miles) in diameter. The distant

Kuiper belt object Eris, roughly the same size as Pluto, which has a diameter of about 2,350 km (1,460 miles), is also designated a dwarf planet rather than a small body, as are Haumea and Makemake, both of which have diameters of about 1,450 km (900 miles). These four dwarf planets are known as plutoids. At least four other Kuiper belt objects are also believed to be larger than Ceres. In the 1990s astronomers recognized that Pluto, which has a diameter of about 2,300 km (1,400 miles), is a giant icy leftover from the time when the planets formed. Pluto had been classified as a planet from its discovery in 1930 but was reclassified as a dwarf planet in 2006. Primarily for historical reasons, the moons of the planets are not considered small bodies. Nevertheless, many moons are asteroid-size or smaller, and some are probably captured asteroids or comet nuclei.

Although small bodies are found throughout the solar system, most of those known are concentrated in several regions. Those regions in which they travel in fairly stable orbits include (1) the main asteroid belt, between the orbits of Mars and Jupiter, (2) the gravitationally stable points (Lagrangian points) along the orbital paths of Jupiter, Mars, Uranus, Neptune, and Earth where objects called Trojan asteroids are clustered, (3) the Kuiper belt, a disk-shaped zone of icy bodies whose members are also called trans-Neptunian objects because they all have orbits beyond that of Neptune, and (4) the spherical Oort cloud of icy bodies inferred to be orbiting the Sun at distances typically more than 1,000 times that of Neptune or Pluto. In none of the above categories is Sedna, an object discovered in 2003 in the outer reaches of the solar system that is perhaps as large as 40 percent the size of Pluto. Sedna has a highly elliptical orbit in which its closest approach to the Sun is more than twice that of Neptune i.e., billions of kilometres beyond the recognized extent of the Kuiper belt; it may be a member of a population of objects lying between the Kuiper belt and the Oort cloud.

Other small bodies travel in unstable orbits that cross the paths of one or more of the planets. Those include (1) most observed comets, (2) near-Earth asteroids, most with orbits that cross either Earth's orbit or Mars's orbit and some with orbits that lie mostly or entirely inside Earth's orbit and cross the orbits of Venus or of both Venus and Mercury while closely approaching Earth's, (3) Centaur objects, icy bodies that are thought to have been gravitationally perturbed out of the Kuiper belt and now travel mainly between the orbits of Jupiter and Neptune, and (4) individual idiosyncratic planet-crossing objects such as the asteroid Hidalgo, which travels between the inner edge of the asteroid belt and a point just beyond the orbit of Saturn. All objects in planet-crossing orbits eventually collide with the Sun or a planet or are permanently ejected from the solar system, although some survive for long periods (up to hundreds of millions of years) because of stabilizing resonances.

Essentially all the small bodies are thought to be remnant material from the planet-building process that took place during the formation of the solar system from the solar nebula. Near the hot nascent Sun, closer than Jupiter's present orbit, only the less-volatile high-temperature-freezing-point substances such as metals and silicates could condense from the nebula. Beginning as tiny grains, that material gradually accreted under mutual gravitational influence to become large rocky bodies up to hundreds of kilometres in diameter. Most of those large objects, called planetesimals, ultimately combined to form the dense rocky planets of the inner solar system (Mercury through Mars). In the outer solar system, near Jupiter's present orbit and beyond, temperatures were cool enough to allow abundant volatile substances such as water and carbon dioxide to condense to their ices. Although rocky planetesimals likewise accreted there, the significant additional mass contributed by the ices allowed the growth of the low-density gas-rich outer planets (Jupiter through Neptune).

Nevertheless, in the inner and the outer solar system, there remained rocky bodies and predominantly icy bodies, respectively, that were never incorporated into planets or captured as planetary satellites. It is now believed that many of these bodies were subsequently mixed together and relocated from their formation locations by the migration of the giant planets. Today, in the inner solar system, these remnants are represented chiefly by the asteroids and their fragments; in the outer solar system, by the objects in and derived from the Kuiper belt and the Oort cloud. Comets originate, and most are still located, in the Kuiper belt and the Oort cloud. Except for the very largest Kuiper belt objects (diameters greater than about 100 km [60 miles]), comets in the Kuiper belt and beyond are unobservable because of their great distances from Earth. It is not until they enter the realm of the planets that they develop comas and tails.

Nice Model

"Nice model" describes dynamical evolution of the outer Solar System since the time when the gas was removed from the protoplanetary disk. In this model, the giant planets underwent a dynamical instability which played a major role in shaping the present-day Solar System. This instability could have happened several 100 million years after the planets formed. The Nice model can explain several observations in the Solar System, including the orbits of the giant planets, the existence and the orbital structure of several small body populations (e.g., Jupiter's Trojans), and the late heavy bombardment.

The best understanding of how the planets have moved in throughout our system's evolution arose from the Nice Model, proposed by an international collaboration of scientists in 2005. This model suggests that at the inner edge of the icy disk, some 35 AU from the Sun, the outermost planet began interacting with icy planetesimals, influencing the second sort of migration to occur: gravitational scattering. Comets were slingshotted from one planet to the next, which gradually caused Uranus, Neptune, Saturn and the belt to migrate outwards. Jupiter's powerful gravity flung the icy objects that reached it into highly elliptical orbits or out of the Solar System entirely, which in order to conserve angular momentum, further propelled its journey inwards.

An extension to this theory is the 'Grand Tack model', which is named after the unusual course of Jupiter's migration towards the Sun before stopping and migrating outwards again, like a sailboat tacking about a buoy. At the distance that Mars would later coalesce, material had been swept away due to Jupiter's presence. This resulted in the stunted growth of Mars and a material-rich region from which the Earth and Venus formed, explaining their respective sizes. The gas giant's travels also prevented the rocky material in the asteroid belt from accreting into larger bodies due to its strong gravitational influence. Although Jupiter swapped positions with the asteroid belt twice the movements were so slow that collisions were minimal, resulting in more of a gentle displacement.

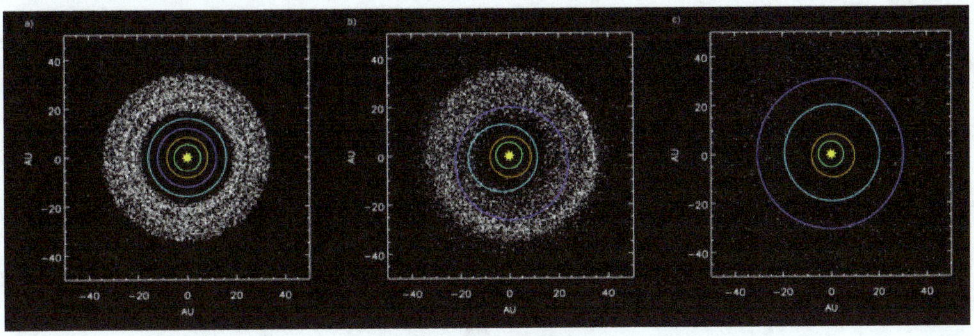

The Nice Model.

But why did Jupiter's migration to the Sun's fiery depths cease? For that it has Saturn to thank. As the two planets moved further away from each other, it was believed they became temporarily locked in a 2:1 orbital resonance. That meant that for every orbit of the Sun Saturn made, Jupiter made two. The Nice Model showed that the planetary coupling increased their orbital eccentricities and rapidly destabilised the entire system. Jupiter forced Saturn outwards, pushing Neptune and Uranus into extremely elliptical orbits where they gravitationally scattered the dense icy disk far into the inner and outer Solar System. This disruption in turn scattered almost the entire primordial disk. Some models also show Neptune to have been propelled past Uranus to become the farthest planet from the Sun as we now know it. Over time the orbits of the outermost planets settled back into the near circular paths we observe today.

The Nice Model explains the present day absence of a dense trans-Neptunian population and the positions of the Kuiper belt and Oort cloud. It also accounts for the mixture of icy and rocky objects in the asteroid belt, like water-rich dwarf planet, Ceres, which likely originated from the icy belt. The rapid scattering of icy objects, around 4 billion years ago, dates with the onset of the late heavy bombardment period, which is predominantly recorded from the Moon's well-preserved surface.

References

- L. Vu (3 October 2006). "Planets Prefer Safe Neighborhoods". Spitzer Science Center. Archived from the original on July 13, 2007. Retrieved 05, August 2020
- Small-bodies-solar-system, focus-areas, solar-system: science.nasa.gov, Retrieved 28 May 2020
- Cassan, A.; Kubas, D.; Beaulieu, J. P.; Dominik, M; et al. (2012). "One or more bound planets per Milky Way star from microlensing observations". Nature. 481 (7380): 167–169. arXiv:1202.0903. Bibcode:2012Natur.481..167C. doi:10.1038/nature10684. PMID 22237108
- Planetary-astronomy, planetary-science-3: planetary-science.org, Retrieved 02 May 2020
- Emspak, Jesse. "Kepler Finds Bizarre Systems". International Business Times. International Business Times Inc. Retrieved 18, February 2020
- Small-body, science: britannica.com, Retrieved 17 July 2020
- Stark, C..; Kuchner, M. (2008). "The Detectability of Exo-Earths and Super-Earths Via Resonant Signatures in Exozodiacal Clouds". The Astrophysical Journal. 686 (1): 637–648. arXiv:0810.2702. Bibcode:2008ApJ...686..637S. doi:10.1086/591442

Permissions

All chapters in this book are published with permission under the Creative Commons Attribution Share Alike License or equivalent. Every chapter published in this book has been scrutinized by our experts. Their significance has been extensively debated. The topics covered herein carry significant information for a comprehensive understanding. They may even be implemented as practical applications or may be referred to as a beginning point for further studies.

We would like to thank the editorial team for lending their expertise to make the book truly unique. They have played a crucial role in the development of this book. Without their invaluable contributions this book wouldn't have been possible. They have made vital efforts to compile up to date information on the varied aspects of this subject to make this book a valuable addition to the collection of many professionals and students.

This book was conceptualized with the vision of imparting up-to-date and integrated information in this field. To ensure the same, a matchless editorial board was set up. Every individual on the board went through rigorous rounds of assessment to prove their worth. After which they invested a large part of their time researching and compiling the most relevant data for our readers.

The editorial board has been involved in producing this book since its inception. They have spent rigorous hours researching and exploring the diverse topics which have resulted in the successful publishing of this book. They have passed on their knowledge of decades through this book. To expedite this challenging task, the publisher supported the team at every step. A small team of assistant editors was also appointed to further simplify the editing procedure and attain best results for the readers.

Apart from the editorial board, the designing team has also invested a significant amount of their time in understanding the subject and creating the most relevant covers. They scrutinized every image to scout for the most suitable representation of the subject and create an appropriate cover for the book.

The publishing team has been an ardent support to the editorial, designing and production team. Their endless efforts to recruit the best for this project, has resulted in the accomplishment of this book. They are a veteran in the field of academics and their pool of knowledge is as vast as their experience in printing. Their expertise and guidance has proved useful at every step. Their uncompromising quality standards have made this book an exceptional effort. Their encouragement from time to time has been an inspiration for everyone.

The publisher and the editorial board hope that this book will prove to be a valuable piece of knowledge for students, practitioners and scholars across the globe.

Index

www.ingramcontent.com/pod-product-compliance
Lightning Source LLC
LaVergne TN
LVHW080716210926
848485LV00003B/476